D0399016

TEDBooks

The Mathematics of Love

Patterns, Proofs, and the Search for the Ultimate Equation

HANNAH FRY

TED Books
Simon & Schuster

New York London Toronto Sydney New Delhi

TED Books

Simon & Schuster, Inc.
1230 Avenue of the Americas
New York, NY 10020

Copyright © 2015 by Dr. Hannah Fry

All rights reserved, including the right
to reproduce this book or portions thereof in
any form whatsoever. For information address
Simon & Schuster Subsidiary Rights Department,
1230 Avenue of the Americas, New York, NY 10020.

TED, the TED logo, and TED Books are trademarks of TED Conferences, LLC.

First TED Books hardcover edition February 2015

TED BOOKS and colophon are registered
trademarks of TED Conferences, LLC

SIMON & SCHUSTER and colophon are
registered trademarks of Simon & Schuster, Inc.

For information about special discounts for bulk
purchases, please contact Simon & Schuster
Special Sales at 1-866-506-1949 or
business@simonandschuster.com.

For information on licensing the TED Talk
that accompanies this book, or other
content partnerships with TED, please contact
TEDBooks@TED.com.

Jacket and interior design by: MGMT. design

Manufactured in the United States of America

10   9   8   7   6   5   4   3   2   1

Library of Congress Cataloging-in-Publication
Data is available.

ISBN 978-1-4767-8488-5
ISBN 978-1-4767-8489-2 (ebook)

CONTENTS

The
Mathematics of Love

I'd like to begin with a confession: I am not an expert in love. I have never taken a course in psychology; I understand only the basics of human biochemistry; and my own dating history—much like everyone else's—is a mixed bag of successes mingled with a healthy series of disasters.

What I am, however, is a mathematician. And in my day job of teasing out and understanding the patterns in human behavior, I've come to realize that mathematics can offer a new way of looking at almost anything—even something as mysterious as love.

My aim in writing this book is not to replace any of the other excellent sources available on the science of human connection. I wouldn't be qualified to describe the intangible thrill, all-consuming passion, or world-ending despair that love can bring. If that's what you're after, might I recommend you simply turn to nearly every painting, poem, sculpture, or song created over the last 5,000 years.

Instead, I want to try and offer you a different perspective on the most talked-about subject in the history of human existence, using mathematics as a guide.

You would be forgiven for thinking that love and mathematics don't seem to naturally sit well together. Human emotions, unlike mathematical equations, are not neatly ordered or well behaved, and the real thrill and essence of romance can't easily be defined.

But that doesn't mean that mathematics doesn't have something to offer. Because mathematics is ultimately the study of patterns—predicting phenomena from the weather to the growth of cities, revealing everything from the laws of the universe to the behavior of subatomic particles. And if we consider them honestly, none of those things is neatly ordered or easily predictable, either.

Thankfully, love—as with most of life—is full of patterns: from the number of sexual partners we have in our lifetime to how we choose who to message on an internet dating website. These patterns twist and turn and warp and evolve just as love does, and are all patterns which mathematics is uniquely placed to describe.

The math will offer a number of dating insights, but I have another confession: The aim of this book isn't just to illuminate your love life. My hope is also to illuminate how beautiful and relevant math is. I wanted to write this book because I'm always a bit disappointed with the way that math is viewed so negatively by the general public, even if I'm not surprised that it has such a bad reputation. Most people's only experience of mathematics is as their most hated subject at school: The topics seemed uninspiring, the ideas hadn't changed in hundreds of years, and the answers were all written in the back of the textbook. It's no wonder some people think math has nothing new to offer. But this just couldn't be further from the truth.

Mathematics is the language of nature. It is the foundation stone upon which every major scientific and technological achievement of the modern era has been built. It is alive, and it is thriving. As the physicist and writer Paul

Davies puts it:

> *No one who is closed off from mathematics can ever grasp the full significance of the natural order that is woven so deeply into the fabric of physical reality.*

To try to convince you of how insightful, relevant, and powerful mathematics can be, I've deliberately tried to choose the one subject that seems as far away from equations and proofs as possible and show how—even in that context—math still has something to offer. I want to share with you my favorite—mathematically verifiable—ways of understanding how love can work.

We'll calculate your chances of finding the person you've been waiting for. I'll show you a mathematical argument to justify approaching someone in a bar. And we'll even perform some mathematical tricks to help you to smoothly plan your wedding.

I've framed most of the examples using the traditional story of man meets woman. This is simply because having two clear groups targeting each other can help to make the math a lot simpler. The choice of examples aside, though, all of the results and tips in the book are general enough to apply to any gender and sexuality.

On occasion we'll use data from real-life couples to offer a strategy for singles in search of someone special. Other times we'll stray into abstraction and oversimplification (as mathematicians so often have a habit of doing) in the hunt for insight. There are elements of economics and science in many of the examples, but the mathematics is always there, even when it's sometimes playing a more subtle role. The examples might not always apply directly to your own love life, but I hope that you will find them interesting regardless.

Most of all, though the goal of this book is to reveal the patterns that govern one of life's greatest mysteries, my great hope is that a little bit of insight into the mathematics of love might just inspire you to have a little bit more love for mathematics.

1 What Are the Chances of Finding Love?

In many ways, we're all the same. Personal quirks aside, few of us would turn down the chance to experience true, romantic love. In one form or another, we are all united by a private quest for lasting happiness. Learning how to attract and keep the partner of your dreams are important aspects of this mission that we'll come to later, but they won't mean anything until you've found that special someone to target with your affections.

For those of us who have been single for any length of time, finding someone special can sometimes feel like an insurmountable challenge. A few years of dating a succession of boring Bernards and psycho Suzys can leave us frustrated and disappointed and feeling like the odds are stacked against us. And some people will tell you that your feelings aren't necessarily unfounded. In fact, in 2010 mathematician and long-standing singleton Peter Backus even calculated that there were more intelligent alien civilizations in the galaxy than potential girlfriends for him to date.

But things might not be as hopeless as they initially appear. There are 7 billion people on Earth, after all, and while not all of them will be to our particular taste, this chapter explains how we can use Backus's method to calculate your chances of bagging yourself a partner—and specifically, why being a bit more open to potential will give you a better chance of finding love on your own planet.

In Backus's paper (titled "Why I Don't Have a Girlfriend"), he adapts a formula used by scientists to consider why Earth hasn't yet been visited by aliens to instead work out how many women would meet his criteria for a girlfriend.

The equation Backus employs was named after its originator, Frank Drake, and aims to estimate the number of intelligent extraterrestrial life-forms in our galaxy. The method is simple: Drake breaks the question down into smaller components, asking about the average rate of star formation in our galaxy, the fraction of those stars that have planets, the fraction of planets that could support life, and the fraction of civilizations that could potentially develop a technology that releases detectable signs of their existence into space.

Drake exploited a trick well known to scientists of breaking down the estimation by making lots of little educated guesses rather than one big one. The result of this trick is an estimate likely to be surprisingly close to the true answer, because the errors in each calculation tend to balance each other out along the way.[1] Depending on the values chosen at each of the steps (and there is some debate over the final few), scientists currently think there are around 10,000 intelligent extraterrestrial civilizations in our galaxy. This is not science fiction: scientists really have convinced themselves that there is life out there.

Of course, just as it's not possible to calculate precisely

1. Breaking the problem down makes the estimate like Brownian motion. An estimate with n steps would have an error that diffused like $\sqrt{n}$.

how many alien life-forms there are, it's also not possible to calculate exactly how many potential partners you may have. But all the same, being able to estimate quantities that you have no hope of verifying is an important skill for any scientist. And the technique—known as Fermi estimation—applies to everything from quantum mechanics to the brain-teaser interview questions used by companies like Google.

It also applies to Peter Backus's quest to see if there are intelligent, socially advanced women of the same species out there for him to date. And the idea is the same: Break the problem into smaller and smaller pieces until it's possible to make an educated guess. These were Backus's criteria:

1. How many women are there who live near me? (In London -> 4 million women)
2. How many are likely to be of the right age range? (20% -> 800,000 women)
3. How many are likely to be single? (50% -> 400,000 women)
4. How many are likely to have a university degree? (26% -> 104,000 women)
5. How many are likely to be attractive? (5% -> 5,200 women)
6. How many are likely to find me attractive? (5% -> 260 women)
7. How many am I likely to get along well with? (10% -> 26 women)

Leaving him with just twenty-six women in the whole world he would be willing to date.

Just to put that into perspective, that means there are

around four hundred times more intelligent civilizations living on other planets than potential partners for Peter Backus.

Personally, I think that Backus is being a little picky. In effect, he's suggesting that he only gets on with one in every ten women he meets, and that he only finds one in twenty attractive enough to go out with. This means he'll have to meet up to two hundred women to find one that fits just these two criteria. And that's not even taking into account whether she likes him.

I think there's room to be a bit more generous. Maybe the numbers should go a little more like this:

1. How many people of the right gender are there who live near me? (In London -> 4 million women)
2. How many are likely to be of the right age range? (20% -> 800,000 women)
3. How many are likely to be single? (50% -> 400,000 women)
4. How many are likely to have a university degree? (26% -> 104,000 women)
5. How many are likely to be attractive? (20% -> 20,800 women)
6. How many are likely to find me attractive? (20% -> 4,160 women)
7. How many am I likely to get along well with? (20% -> 832 women)

Almost a thousand potential partners across a city, then. Seems much more like it in my book.[2]

---

2. And this *is* my book.

But there is another issue.

If Backus could relax some of his criteria just a bit, he'd have a much bigger pool of potential partners to work with. In fact, he could instantly quadruple his chances if he were a little less fussy about his future love holding a university degree. And the pool of ladies would be much, much larger if he were willing to expand his search to outside of London.

Strangely, though, opening our minds to all potential partners seems to be the opposite of what we do when we're single. I recently heard of a gentleman with an even clearer idea of what he was looking for in a potential partner. This man had set up a profile on the dating website OkCupid, which offers a profile section where you can outline certain "deal-breakers": things that you can't tolerate under any circumstances. His list ran to over a hundred, and was so extreme that it became the subject of a popular article on the website BuzzFeed. Under the heading "Do Not Message Me If" were the following gems.

1. You needlessly kill spiders
2. You have tattoos you can't see without a mirror
3. You discuss Facebook in the visceral world
4. You consider yourself a happy person
5. You think world peace is actually a goal of some sort

As reasonable as it is to limit your search to a spider-loving, ink-free peace hater, unfortunately, the more deal-breakers you have the less likely you are to find love. Because when you feed a mammoth list like this one into

Backus's equation—or even my version—unfortunately, you'll get an answer close to zero potential partners.

Of course we all have must-haves and no-nos when it comes to love. But an extensive list like this does raise an interesting question. Just how much do our preemptive dating criteria actually harm our chances of finding love?

The reality is that when people are single and looking for a prospective partner, they often add in all sorts of must-haves or must-not-haves that dramatically reduce their chances. I have a very close friend who ended a potentially fruitful courtship simply because the gentleman wore black shoes with blue jeans to a date. I have another chum who insists that he cannot date a woman who uses exclamation marks! (That one is for him.) And how many friends do we all know who will not consider someone unless they are driven enough, or gorgeous enough, or rich enough?

Being good on paper doesn't mean anything in the long run. There's no point in restricting your search to people who match everything on your checklist, because you're just setting yourself an impossible challenge. Instead, pick a couple of things that are really important and then give people a chance. You might just be pleasantly surprised.

Let's be honest, we probably all know people who've ended up with someone they never thought they'd be with, even if that person were the last life-form on the planet. After all, in the words of Auntie Mame, "Life's a banquet, and most poor suckers are starving to death!"

Just ask Peter Backus. He beat his own odds; he got married last year.

2 How Important Is Beauty?

If Peter Backus's story has persuaded you to be a bit more flexible in your criteria, your next step is to figure out how to attract the object of your desire.

Choosing a partner is one of the most important decisions you'll make in your lifetime—so much of your future happiness rests on whom you pick to settle down with. And there are surely a number of things that we'd all like to find in a partner: a willingness to compromise; an ability to provide for you and your family; someone who is warm, forgiving, and supportive. But if these are the really important traits, have you ever wondered why we are all so obsessed with how hot a person is?

Plump lips and big biceps might be nice to look at now, but they won't be much help at four o'clock in the morning when your baby's diaper needs changing, or in sixty years' time when your catheter bag needs replacing. And yet we have been preoccupied by beauty since the dawn of civilization. Is it possible that people in every society have tricked themselves into thinking that something as frivolous and transient as beauty is what's most important? Or, given how pervasive the theme of beauty has been throughout history, perhaps there is something more subtle at play.

Scientists, mathematicians, and psychologists have dedicated considerable brainpower over the centuries to the pursuit of defining the elusive essence of beauty. Although many of these ideas are based more in science than mathematics, it's worth knowing what you're up against in the fight for affection and why beauty does go more than skin deep. But I wouldn't quite advocate running out and buying yourself a new face just yet—later in the chapter we'll also look into how we can exploit the rules of human perception to make ourselves more attractive without going under the knife.

A universal rule of beauty

Hot-or-not debates are only interesting because people see beauty differently. But there are a lucky few—mainly in Hollywood—whose faces are so beautiful they seem to be able to unite opinion on their looks. There must be some basic criteria that we all agree on. And if we all subconsciously understand these rules, it should be easy to define what it is that makes those faces stand out.

Some people believe that we've already arrived at the definitive answer to what makes a person beautiful, and claim that it lies in a mathematical concept called the "golden ratio."

If you haven't heard of the golden ratio, it's an irrational number approximately equal to 1.61803399 . . . and usually denoted by the Greek letter *phi*, or Φ. Its definition comes from geometry, but it has been found to apply in everything from the number of petals on a flower to the population growth of rabbits.

It has also repeatedly been linked to human beauty.

You may have heard it said that the perfect face should have a mouth that is 1.618 . . . times larger than the base of the nose, eyebrows that are 1.618 . . . times wider than the eyes, and so on.

On the face of it, this might seem like it makes sense. Eyes that are too far apart or practically touching might not fit most people's definition of beauty. And applying the golden ratio to human faces does produce seemingly persuasive results. Dr. Stephen Marquardt, a plastic surgeon, has even developed a golden ratio mask that he uses as a guide to help him design surgical intervention for his more facially challenged patients. The mask has been overlaid on the faces of famously beautiful women like Angelina Jolie and Elizabeth Taylor, all of whom were seen to have characteristics which conformed to the mask.

Linking beauty to the golden ratio is a neat theory that you'll find espoused in numerous beauty blogs and YouTube videos. There's just one problem—it isn't good science.

Real science is about trying as hard as you can to *disprove* your own theories. The more you try, and fail, to prove yourself wrong, the more evidence there is to suggest that what you're saying is right. As much as I'd like beauty to be defined by a single number, I'm afraid that trawling through thousands of faces and measuring every possible ratio until you find something that fits your theory just isn't science.

The problem with using the golden ratio to define human beauty is that if you're looking hard enough for a pattern, you'll almost certainly find one, especially if you're prepared to be a little loose with your definitions.

How do you decide where the "start" of your ear is, or the point at which your nose definitively "ends"? And how do you do this to a degree of accuracy of five or more decimal places in your golden ratio measurement?

Perhaps there will come a time when somebody does find a good reason why the human body has a thing for this number. But until then, as Stanford University mathematician Keith Devlin puts it, the golden ratio as a definition for beauty really does seem to be "the myth that won't go away."

But, thankfully for the objectives of this book, there *are* some mathematical ideas that really do seem to be linked to beauty. And each has its own explanation of why we are evolutionarily preprogrammed to prize certain characteristics in a potential partner.

One of the first to be discovered was our preference for an average face shape. Since the 1800s, researchers have known that overlaying images of lots of faces from a particular ethnic group will lead to an average face that is widely considered attractive. Each ethnic group has a different ideal, but essentially, once you've ironed out the oversized chins, lopsided ears, and lengthy foreheads, what you're left with is a completely average (if unexciting) hottie.

The theory is that when looking for a partner, we tend to dislike unusual face shapes for fear that they mask a weird genetic mutation that we'd like to avoid passing on to our future offspring.

Thoughts of the health and success of our future children are a recurring theme when judging for beauty. Facial symmetry, too, stands out as an important factor for beauty, and people with naturally symmetrical faces consistently score highly on attractiveness surveys. But it seems that when picking out symmetrical faces as beautiful,[1] we're doing nothing more than validating an underlying clean bill of health.

Whenever you pick up a cough or cold during your childhood it will have a tiny impact on your development, leading to slightly unusual patterns of growth. One eye might end up just a few millimeters higher than the other, or one nostril ever so slightly larger than the other. The effect might be small, but it's enough, it seems, for people to subconsciously pick up on these cues when judging beauty. On some subliminal level, we all know that someone with slightly mismatched features probably doesn't have the best immune system. After all, you want your future offspring to be as healthy as possible.

And evolutionary influence on our thinking about beauty doesn't end there. It also comes into play in the characteristics in men and women that we are universally drawn to. Female faces with narrow chins, large eyes, and fuller lower lips are rated highest in beauty across many different cultures. Likewise, there is a broad preference for male faces with strong brows and well-defined jaws. The significance of these traits seems to lie in their connection to the prevalence of male and female hormones.

As girls go through puberty, their hormones will have a direct impact on how their facial features develop. Women with high levels of estrogen will end up with full lips and

1. Just so we're clear, we're talking about reflection symmetry here. Rotational symmetry in a face is generally considered a bad thing.

a large waist-to-hip ratio, while women with lower levels of androgen, the steroid hormones, will keep their short and narrow jaws from childhood, along with their flatter brows—giving them much larger eyes.

And—surprise, surprise—this balance of female hormones is also positively linked to fertility.

Men, on the other hand, need testosterone throughout puberty to develop muscle mass, broader jaws, and defined brow ridges, which will inevitably result in more sunken eyes. And testosterone, the male sex hormone, is also a useful marker for fertility.

So all we're really doing when picking a guy with a strong jaw, or a woman with beautiful plump lips, is giving in to our evolutionary desire for offspring. That's why women wear lipstick. It's so you want their babies.

Personal preferences

But there's no need to run to the plastic surgeon just yet. Even with all these seemingly universal rules, there's still an awful lot of room for personal preference. And for everything we've said about symmetry and hormones, sometimes it's the people who break the rules who come out being rated as the most attractive.

For example, it seems that the rules of symmetry only really work on *pictures* of people's faces. In real life, many people are also drawn to asymmetrical characteristics. Not only do they show more character, but people who possess them are also perceived to be more sincere. During speech, 76 percent of us will have a more pronounced movement on the right-hand side of our mouths. You might not notice it if you're not looking for it, but it seems that subconsciously people find asymmetry in expression much more natural, and hence, more attractive.

Likewise, it's not true that the more masculine or feminine your face is, the better. We all have certain personality traits we find particularly attractive in a partner, and—although it's not quite physiognomy—there is now evidence to suggest that the qualities we desire in a mate are reflected in our preference for different faces.

For example, testosterone, which causes those square jaws and defined brows, is also the hormone that causes people to be assertive and aggressive. But some women prefer a more laid-back partner. Likewise, for the men who prefer a more aggressive woman, those big eyes and small chins might just make them look a bit too vulnerable. Some people will prefer a partner with more bite.

You might be surprised at how easily you can perceive personality traits from a face. Most of us can do it without even realizing. Take the pictures below, for instance: Which male and female faces look more assertive? And which look more easygoing?

If you chose B and D as the more assertive, then you match the responses of 90 percent of the population. These images are constructed from overlaying the faces that people who highly value assertiveness found most attractive. Likewise, A and C are made from overlaying faces considered attractive to people who prize an easygoing partner. Similar results are found for other characteristics, too: People who are looking for an extrovert find faces that people can spot as extroverted to be attractive; the same goes for people who desire introverts or neurotics. Beauty truly is in the eye of the beholder.

There is a lot more to the science of attractiveness,[2] but ultimately, beauty is what lies between the equations. Everyone truly has a unique ideal, so there's no mathematical solution here. All this means there's really no point stressing out about it. Focus more on developing some amazing small talk and killer charm.

Changing how people see you

So, perhaps it's not possible to change your face to make it more universally attractive. But when it comes to picking a partner, the role of human choice comes into play. Choice means probability, and probability means mathematicians can get to work.

When someone decides to approach you in a bar, or to accept your advances at a party, they're not judging your beauty against every other face in the world. Nobody minds that you don't look like George Clooney or Heidi

---

2. See, for example, *In Your Face* by David Perrett for a well-written and comprehensive overview.

Klum. They're making a decision based on the options available to them at the time, and this is perhaps where there's room to use a mathematical idea to your advantage.

By defining these options in equations, we can create a language to explain why we all make the choices we do, in what has become known as "discrete choice theory."

Despite our illusions of free will, there are some simple rules that people will often follow when coming to a decision. These rules mean that people's choices are surprisingly easy to manipulate. As the economist Dan Ariely puts it, we're all just a little bit "predictably irrational."

Imagine that you're in a movie theater and choosing which snacks to buy. Perhaps a small popcorn costs $5.00, while the large popcorn is an eye-watering $8.50. The large option seems terribly expensive, until the cashier points out that the large popcorn is only $.50 more than the medium. No sensible person would ever choose to buy the medium popcorn when you could have the large for only a few cents more, but the fact that the medium popcorn is on the menu has a big impact on your decisions: it serves to make the large popcorn look like a much better deal.

This is known in economics as the "decoy effect." What it demonstrates is that the presence of an irrelevant alternative can change how you view your choices. It has been exploited by marketing experts for decades. But it also has the potential to help make you seem more attractive.

In his 2008 book, Dan Ariely explains the impact of the decoy effect on perceptions of human beauty.

By surveying his students regarding the attractiveness of a range of male faces, Ariely found two that were considered equally attractive; we'll call these two men Adam and

CONDITION A

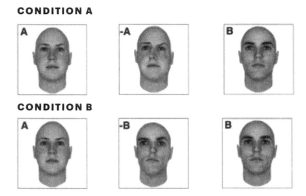

CONDITION B

Ben. Using Photoshop, Ariely created uglified versions of both Adam and Ben and then made two sheets of faces to test his theory.

The first sheet depicted Adam and Ben as normal, but included the uglified version of Adam, as in the first row of the image above. The second sheet also contained the original pictures of Adam and Ben's faces, but instead included the ugly version of Ben, as in the second row.

He handed out these sheets to six hundred of his students, with 50 percent of them viewing the first sheet and 50 percent viewing the second, and asked each participant to say which face they considered most attractive.

No one chose the uglified images, but their presence on the sheets had a dramatic impact.

Where ugly Adam appeared, 75 percent of participants said they found the original Adam the most attractive. Where ugly Ben appeared, the exact opposite happened: 75 percent of people thought that the original image of Ben was the best-looking.

On both sheets, the uglified versions of Adam and Ben served to make the original images more appealing, just as the decoy effect would predict.

The message for making yourself appear more attractive is clear. When going to a party to talk to potential partners, choose a friend to go with you who is as similar-looking to you as possible, except slightly less attractive. Having them there will make you seem like a better option.

If this seems callous, remember that making these judgments is something we all do instinctively. Math is the language of nature, and by listening to what the mathematics tells us, we can all gain a better understanding of how and why we do the things we do.

After all, as George Bernard Shaw put it, "Love consists of overestimating the difference between one woman and another." So don't be shy to use the decoy effect to your advantage.

3 How to Maximize a Night on the Town

Although most of this book is dedicated to the search for true, lasting, romantic love, from time to time, women and men have both been known to operate on a much baser level. For some, a Friday night out isn't complete without ending up in a stranger's bedroom. For others, a sweaty make-out session on the dance floor is enough to make their evening. Whatever your ambitions are, this chapter will show you just how to maximize your chances with the object of your affection, or at least the next best thing, on a night out.

Say you're at a party with a group of single friends, all trying to decide how best to boost your chances of meeting someone. Should you sit back and wait for them to come to you, or walk right up to the prettiest partygoer, risking a humiliating rejection? And who should you approach to give you the best chance of success?

If we all go for the blonde

Anybody who has seen the 2001 film *A Beautiful Mind* might think that math already has the answer. The film follows the life of mathematics superstar John Nash and includes some dramatized explanations of his major mathematical breakthroughs. In one famous scene, Nash and his three charming gentleman friends spot a group of five women in a bar: four brunettes and one particularly beautiful blonde.

All of the men are immediately drawn to the blonde. But, rather than all rushing to shower her with attention, Nash argues for a different tactic. Strategically, he suggests they would all do better by ignoring the blonde and aiming for her four brunette friends instead:

> If we all go for the blonde, we block each other and not a single one of us is going to get her. So then we go for her friends, but they will all give us the cold shoulder because nobody likes to be second choice. But what if no one goes to the blonde? We don't get in each other's way and we don't insult the other girls. That's the only way we win.

Let me just pause to point out the implicit assumptions here:

1. The blonde will hook up with anyone, as long as only one man approaches her.
2. The women deserve no say whatsoever in the way partners are divided.
3. Everyone would rather get with someone they didn't really find very appealing than not have any prospects at all.

That delightful display of 1950s equality aside, the example does illustrate an interesting and counterintuitive point: it's not always best to go for the person you are the most attracted to. In this scenario, at least, everyone ends up better off if they ignore their personal preferences.

The mathematics hidden behind this problem is game

theory—a way to formalize strategies and find the best tactic in a situation.

Despite the name, game theory isn't just about studying activities undertaken for amusement. It can apply in any situation where two or more adversaries compete against one another for some sort of payoff. Here, the friends are competing to win the women, but the theory has been successfully applied to everything from evolutionary biology (animals with different traits within a species compete for food or resources) to economics and politics (governments balance competing interests to influence the behavior of their citizens).

In the example from *A Beautiful Mind*, one way that all the men can win is indeed to ignore the blonde. However, fictional Nash's plan has a flaw: Each of the men could easily dupe his chums into thinking he was following the plan, but at the last minute switch from his brunette, approach the blonde, and end up the triumphant winner. All the men would still leave with a girl, though this is generally a dangerous path to go down if you'd like to keep your friends.

It wouldn't be sensible to stab your friends in the back if Nash's first assumption had been wrong, though. If the blonde had an obvious preference for the best-looking man and showed no interest in the other three, then the strategies for everyone would be clear. The best-looking man should go for the blonde, while the other three should pair off with the brunettes. In that case, if any of the three tried to switch to the blonde at the last minute, their attempts would be rejected and only damage their chances with the brunettes.

All the men would then be doing what's right for themselves (this is called a "Nash equilibrium"), and at the same time doing what's best for the group (making this also a "Pareto equilibrium").

Sadly, it's rather rare to find such a neat real-world situation, with four opinion-free brunette clones and one standout blond babe whom everyone is madly in love with. In real life, people have different preferences in a group, and generally it's difficult to persuade people to ignore those preferences for the greater good.

So maybe we have to leave game theory behind for the time being. But that doesn't mean that math doesn't have any tips to help you on your night out. To give us a more realistic view, we can turn our attention to a rather neat theory that sheds light on how forward you should be on a night out.

Who to approach at a party

Imagine that three boys start chatting with three girls at a party. Picking some names completely at random, we'll call our six singles Joey, Chandler, Ross, Phoebe, Monica, and Rachel. Let's assume that each of the boys and girls has their own personal preferences; an ordered list of who they'd like to hook up with.

Even though all characters and events in this scenario are entirely fictional and bear no relation to any heavily copyrighted TV series, I've decided—at random—to make Monica and Ross brother and sister. But I've also decided that they'd rather leave the party together (platonically) than alone, and so each appears as the other's third choice.

Rachel is the most popular of the girls, appearing first

	1ST CHOICE	2ND CHOICE	3RD CHOICE
Ross	Rachel	Phoebe	Monica
Chandler	Rachel	Monica	Phoebe
Joey	Phoebe	Rachel	Monica

	1ST CHOICE	2ND CHOICE	3RD CHOICE
Rachel	Joey	Ross	Chandler
Phoebe	Ross	Chandler	Joey
Monica	Joey	Chandler	Ross

on both Ross and Chandler's lists. Meanwhile, both Rachel and Monica have ranked Joey at the top of their list of boys. These clashes in people's desired outcomes mean that—if everyone is to leave with a partner—there's going to have to be some sort of compromise.

If we allowed this scenario to play out in the very old-fashioned boy-approaches-girl way, each of the boys would hit on his first-choice girl.

Since Rachel is hit on by both Ross and Chandler, she gets to pick between them. Ross appears higher on her list of preferences, and so Rachel and Ross hook up—at least for the time being, as Rachel is secretly hoping that Joey will pay her some attention.

Chandler, now single and on the hunt, goes for his second-choice girl—Monica. Since Monica has had no other offers, she hooks up with Chandler, but like Rachel is secretly hoping Joey will hit on her.

Phoebe, without an offer from Ross or Chandler, hooks up with Joey.

So it's settled. All the boys have gotten themselves a girl, and the couples are:

1.   Ross—Rachel
2.   Chandler—Monica
3.   Joey—Phoebe

We now find ourselves in a situation where none of the boys could improve. Only Chandler is without his top choice—Rachel—but she already rejected him. The boys have no incentive to switch partners, even if the girls now decided to approach their preferred boys. Rachel might well prefer Joey, but Joey has paired off with his first choice and would show no interest in switching.

The situation is less appealing for the girls. Rachel, Phoebe, and Monica end up with their second, third, and second choices respectively. Not great in a list of three, and especially when compared to the boys, who ended up with their first, second, and first choices respectively.

This setup is known as the "stable marriage problem," and the process through which the friends picked their partners is called the Gale-Shapley algorithm. If we look into the math behind these couplings, some extraordinary results appear. Regardless of how many boys and girls there are, it turns out that whenever the boys do the approaching, there are four outcomes which will always be true:

1.   Everyone will find a partner.
2.   Once all partners are determined, no man and woman in different couples could both improve their happiness by running off together (for example, Phoebe might still have eyes for Ross, but he's happy with Rachel).

3. Once all partners are determined, every man will have the best partner available to him.
4. Once all partners are determined, every woman will end up with the least bad of all the men who approach her.

The last two points demonstrate a particularly surprising result: In short, the group who do the asking and risk continual rejection actually end up far better off than the group who sit back and accept a suitor's advances.

We can redo our simple example by swapping the male and female roles to see this principle in action. Having the girls approach the boys, and using the same process, would result in the following pairings:

1. Rachel—Joey
2. Phoebe—Ross
3. Monica—Chandler

Now the girls achieve their first, first, and second choices: a marked improvement. This time, the boys only manage their second, second, and second choices: a worse outcome than when they did the asking.

This result does make some intuitive sense. If you put yourself out there, start at the top of the list, and work your way down, you'll always end up with the best possible person who'll have you. If you sit around and wait for people to talk to you, you'll end up with the least bad person who approaches you. Regardless of the type of relationship you're after, it pays to take the initiative.

The difference in outcomes between those who do the asking and those who wait to be asked is particularly important when the stable marriage problem is applied beyond imaginary couples at a party: something the US government found out the hard way.

Through the National Resident Matching Program, the US government has been using the Gale-Shapley algorithm to match doctors to hospitals since the 1950s. Initially, the hospitals did the "proposing." This gave the hospitals the students they wanted, but didn't work well for doctors who had to move halfway across the country to accept their least bad offer. It meant the system ended up full of unhappy doctors and, hence, unhappy hospitals. The organizers gave doctors the role of proposer when they found that out.

But it's not just hospitals and Friday-night action. The Gale-Shapley matching algorithm has been exploited in a host of real-world scenarios: dental residencies, placement of Canadian lawyers, assignment of students to high schools, and the sorority rush. It's so useful that there is a huge amount of academic literature dedicated to investigating a range of extensions and special cases—many of which still apply to the original dating problem.

Mathematicians have adapted the method to allow both men and women to approach either gender simultaneously, and changed the rules to include ties in preference lists, or scenarios where you'd rather go home alone than hook up with the weird guy in the corner. Academics have even explored what happens when you have cheating men (not cheating women, though, strangely).

The math in these special cases can get quite heavy in places (although there are lots of lovely references at the end of this book if you're interested in finding out more). But for all the extensions and examples, the message remains the same: If you can handle the occasional cringe-inducing rejection, ultimately, taking the initiative will see you rewarded. It is always better to do the approaching than to sit back and wait for people to come to you. So aim high, and aim frequently: The math says so.

4 Online Dating

So hopefully you're now bold enough to approach the hotties at a party, armed only with your knowledge of the stable marriage problem. But too many parties in a row can be a bit exhausting, and not many of them will have a Joey or a Rachel to keep things interesting. So why not turn to an approach that can bring you success from the comfort of your living room? It's time for online dating.

At this point, almost everyone knows a couple who met on an internet dating site. In spite of the old stigmas, we've embraced this new approach to finding love in a big way. The latest statistics suggest that three-quarters of US singles have tried dating sites, and up to a third of newly married couples originally met online.

The appeal is obvious. There's no need to pluck up the courage to approach a girl in a bar while in front of an audience of her friends and your friends, or go through the mortification of being set up with the charismatic equivalent of a house plant. With the internet, there really are plenty more fish in the sea. Today's dating sites mean easy access to countless singles tailored to your exact desires, with your perfect match only a click of a button away.

Or at least that's how we think it should be—but sometimes too much choice can just make it harder to weed out the bad options. For some of us, online dating is a succession of frogs with no prince at the end of it. For others, more choice seems to equal more rejection. The good news, as ever, is that math can help.

For mathematicians like myself who study the patterns in human behavior, online dating is the gift that keeps giving. The footprints that people leave online have led to a number of fascinating new insights into love. By experimenting on unsuspecting singles, mathematicians are increasingly able to craft a scientific approach to matchmaking. From following the relationships of people who met through dating sites, we also now know why all existing attempts at scientific matchmaking don't really work—at least, not in the way we want them to. And through looking at the types of people who are popular online, scientists can offer advice on how to make yourself stand out in the increasingly populated crowd of internet singles.

I could have written a whole book on online dating and what it can teach us about ourselves. Sadly, you'll have to make do with just this chapter, but hopefully it will still give you a hint as to how math can help our modern search for love.

How to calculate a matchmaking statistic

Online dating websites are the perfect easy-access catalog of datable strangers, allowing you to filter for age ranges and locations to kick off your search. But if you're looking for something more specific, some websites go even further, offering users a scientific approach to matchmaking.

These websites sift out singles who don't conform to your ideals, as well as suggesting partners you might otherwise overlook if your search only took looks or location into account. One of the most successful websites with this approach is OkCupid, a free dating website founded by a group of mathematicians, with a particularly elegant algorithm at its heart.

An algorithm is similar to a recipe: a series of logical steps that can be used to perform a task. In this case, the OkCupid algorithm takes the questionnaire members fill in on joining and, through a series of logical steps, it generates a score between couples to illustrate how good a match they are.

The three key ingredients are 1) your answers, 2) the answers you'd like a partner to give, and 3) how important each question is to you.

The last of these ingredients is particularly important, as it allows each singleton to personalize the process. Perhaps your future partner's political affiliation is more important to you than whether or not they want to have children, or maybe the opposite is true. Maybe the amount that somebody earns, or how much they like Ryan Gosling films, is a mandatory criterion in a prospective relationship (although if that's the case, you may wish to reconsider; see chapter 1). Each person needs a mechanism to filter out what actually matters.

By asking "How important is this question?" the OkCupid team can assign a value to your possible answers:

1. Not at all important 1
2. A little important 10

3. Somewhat important 50
4. Very important 100
5. Mandatory 250

These values then set the maximum score that your potential match can earn on any question.

To show you how the algorithm works to calculate your match percentage, let's use an example formed by picking two names completely at random: "Harry" and "Hermione."

This example has just two questions: "Do you like quidditch?" and "Are you good at defeating dark wizards?"

Harry	DO YOU LIKE QUIDDITCH?	ARE YOU GOOD AT DEFEATING DARK WIZARDS?
Harry's answer	Yes	Yes
The answer Harry is looking for from his potential partner	Yes	Yes
Importance of the question to Harry	A little important	Very important

Hermione	DO YOU LIKE QUIDDITCH?	ARE YOU GOOD AT DEFEATING DARK WIZARDS?
Hermione's answer	Yes	No
The answer Hermione is looking for from her potential partner	No	Yes
Importance of the question to Hermione	Not important	Mandatory

Using these answers, the algorithm used to calculate how good a match Harry and Hermione are for each other can then be split into a simple three-step process:

Step One

First, we need to calculate how good a match Hermione is for Harry.

Harry has rated the first question as only "A little important," meaning Hermione can score a maximum of 10 points. Since her answer matches what Harry is looking for, Hermione gets 10 out of 10 possible points for the first question from Harry's perspective.

Harry rates the second question as "Very important," so Hermione, having answered "No," gets zero on this question. Her total match percentage for Harry is then $(10+0)/(10+100) = 10/110 = 9.1$ percent.

Step Two

Next, we repeat the first step, except this time calculating how good a match Harry is for Hermione.

Looking at Hermione's preferences, the first question is worth 1 point to her, since she rates it as "Not important." Since Harry answered "Yes," and she was looking for a "No," Harry fails to score. Perhaps Hermione doesn't want someone who talks about quidditch all the time (something we can all relate to).

Meanwhile, the second question is worth a whopping 250 points to Hermione, and let's face it, who doesn't get all hot and steamy over a well-timed Expelliarmus Charm? Harry scores the full 250.

Harry's total match percentage for Hermione then is (0+250)/(1+250) = 250/251 = 99.6%. Hermione just *cannot* get enough of Harry.

Step Three

The final step in the process is to combine the two scores to achieve an overall match.

Many people would naturally reach for the arithmetic mean when asked to calculate an average. The equation has been burned into the backs of most of our eyelids from our school days, but for those who've managed to forget the formula, we'd add Hermione's 99.6 percent compatibility with Harry's 9.1 percent compatibility and divide it by two to get 54.35 percent, a value which is 45.25 percent away from both Harry and Hermione's original match percentages.

When it comes to dating, though, both people's opinions are important. A date where one person has the time of their life while their date is counting down the minutes until they can go home is quite different from one where both people have a reasonably good time. But both scenarios could feasibly end up with an arithmetic mean of 54.35 percent. We need to use a different type of average if we want to distinguish between the two scenarios.

A more sensible average to use in this case is the geometric mean, which is based on multiplication rather than addition. With just two questions, as we have here,[1] the formula to calculate the overall match is:

1. For n questions, the formula becomes: $\left(\Pi_{i=1}^{n} a_i\right)^{\frac{1}{n}}$

*(Hermione's match percentage × Harry's match
percentage)^(½)*

Or $(99.6 × 9.1)^{½} = 30.1\%$ match.

By multiplying the values rather than adding them,
the geometric mean finds a number in the multiplicative
middle (30.1 percent is 3.3 times bigger than 9.1 percent
and 3.3 times smaller than 99.6 percent) and offers a much
fairer way to take both people's opinions into account.
Harry might have ticked all Hermione's boxes, but Harry
would have been infuriated by Hermione's lack of dueling
prowess—hence the 30.1 percent compatibility.

And that's it—apply this algorithm to the hundreds of
available questions and repeat for each of the millions of
users on OkCupid and you've got everything you need for
one of the world's most successful dating websites. It's one
of the most elegant approaches ever attempted to pairing
couples based on their personal preferences. Together
with eHarmony and other similar websites, OkCupid sits
alongside Amazon and Netflix as one of the most widely
used recommendation engines on the internet.

But there's one problem—if the internet is the ultimate
matchmaker, why are people still going on terrible dates? If
the science is so good, surely that first date will be the last
first date of your life? Shouldn't the algorithm be able to
deliver the perfect partner and leave it at that? Maybe the
questionnaires and match percentages aren't all they're
cracked up to be.

Accounting for chemistry

I once went on an internet date where the young gentleman saw fit to steal my shoe halfway through the meal. On
another, I came back from the bathroom to find my date
had put on my jumper and ripped it in the process. It didn't
seem to matter how detailed my online profile was or how
many questions I answered on the website, I still routinely
found myself seated opposite yet another person asking if
my red hair tasted like strawberries.[2]

Personal preferences and individualized lists are the
ideal ingredients to filter our searches according to our
own criteria. But eighty-odd years of relationship science
has taught us one important thing: Trying to use *individual* data to predict how well a *couple* will get along just does
not work.

The problem is that we don't really know what we want
until we find it. Unlike with Amazon or Netflix, where we
truly know our tastes in films and other products, a questionnaire about our personal preferences just isn't enough
to predict who will make us happy. Ultimately, finding a
partner is just a lot more complicated than buying a DVD
box set.

You and I may both really like watching Ryan Gosling
films, but that says nothing about whether we would enjoy
watching them together. And while a mutual respect for
Ryan Gosling may be a good place to start an early conversation or a first date, it's unlikely to be an important predictor of how compatible we'll be in a long-term relationship.

But it's not just trivial measures like film preferences

2. This is all true.

that fail to capture our chances of success as a couple. It's every possible combination of personalized data: demographics, political persuasions, family ambitions, etcetera. None of these can offer a significant or meaningful measure of how compatible you'll be with a prospective partner in real life.

OkCupid even acknowledged that match scores have achieved limited success in finding "longer-term" compatibility in their brilliantly titled blog post "We Experiment on Human Beings!" To test the effectiveness of the company's matching algorithm, programmers instructed the computer to lie to a select group of users, telling them they had a 90 percent match with another member, when in reality the pair were only 30 percent compatible according to the OkCupid algorithm.

The experiment had some interesting results: The probability that singles would send each other an initial message rose from 12.4 percent to 14.5 percent when they were duped into believing they had a higher match score.

People, then, are much more likely to engage in conversation if they are told they are a good match, suggesting that OkCupid's users put some faith in the algorithms. Perhaps that's not too much of a surprise, but you might think that the conversations would fizzle out pretty quickly once the pair realized they weren't compatible.

It turns out that most of them did. Once the first message had been sent, just 15 percent of the users who had been lied to engaged in a conversation of four messages or more—though that's still quite a jump from the equivalent figure of 9 percent for users who knew they were incompatible.

But although 15 percent of the duped-but-ill-suited couples engaged in conversation with each other after initial contact, the figure for people who really were a 90 percent match (and not just told they were) was a surprisingly similar 17 percent. The well-matched couples didn't really get along any better with each other.

The tiny margins between these two numbers mean the OkCupid matching algorithm has its limits in being able to predict the real success of a match. Of course it's easier to engage in a conversation if you have more in common, but only just. And it won't necessarily help you in the long run.

This isn't a flaw in OkCupid's science. Their algorithm is doing exactly what it was designed to do: deliver singles who meet your specifications. The problem here is that you don't really know what you want. So an algorithm that can accurately predict your compatibility with another person simply does not exist yet.

But we might not be too far away. Because although our minds might not be able to tell us what we want, our bodies certainly know it when we see it.

Anyone who has met someone they immediately click with can tell you how exciting it feels, but they may not be aware of how our actions subtly shift to give telltale signs of the connection. Scientists have known for a long time that our body language will mirror that of someone we are attracted to. Our pupils will become dilated, the words we use in conversation will adjust to mimic the language patterns of the other person, and our laughter will begin to synchronize. All of this happens within a matter of minutes, and all of these signs can be used to quantitatively define a connection between two people.

But—perhaps surprisingly—the signals we give off when we first meet someone have also been shown to link to longer-term compatibility within a couple, and can offer a much more reliable indicator than anything that can be derived from a questionnaire.

Eli Finkel, a professor of psychology at Northwestern University, has done a lot of work on the so-called "non-conscious synchrony" that happens between two people, and believes that much of the technology to integrate these measures into online matchmaking already exists—or is just around the corner.

Imagine if you could have a series of short online speed dates on a Skype or FaceTime-like system over the course of an evening. Siri-type technology could track your language patterns, while image recognition software could keep a log of your body language. At the end of your evening, a realistic and meaningful compatibility statistic for your matches could be delivered, giving you a much better basis on which to judge who is worthy of being graced with your real-life presence.

And mathematics, as the language of science, will play a pivotal role in every one of these developments.

It's an exciting prospect, but I think these ideas are more likely to enhance the existing matchmaking algorithms than to replace them. There will always be demand for a range of ways to search for dates, from the detailed and personalized but time-consuming algorithms to the low-effort services like Tinder and Grindr. No website will ever be able to guarantee perfect matches every time, but you can find someone right for you if you're prepared to put in the work.

Picture-perfect

Another problem with matchmaking through questionnaires is the widely held suspicion that it's all about the
photo. In fact, dating services like Tinder and Grindr
have dispensed with the detailed "About Me" section
altogether, instead allowing you to flick through images
of singles in your area and base your dating decisions
on looks alone. But the majority of us who don't fit a
prescribed notion of beauty may be pleased to hear that the
digital world isn't quite as judgmental as you might think.

For the past decade or so, math major and OkCupid
cofounder Christian Rudder has been collecting data on
the users of his website to study the way that people behave
on dating sites. He has come up with some fascinating
findings on topics from understanding how we speak about
ourselves when looking for love to what we say and how we
interact with one another in the earliest stages of romantic connection, as well as some surprising data about the
importance of attractiveness.

My personal favorite finding is that it turns out how
good-looking you are does not dictate how popular you are
on an online dating website—actually, having some people
think that you are ugly can work in your favor.

In one of the voluntary sections of OkCupid, you can
rate how attractive other people are on a scale between
1 and 5. To try and test how attractiveness might link
to popularity, the OkCupid team took a random selec-
tion of 5,000 female users and compared the average
attractiveness scores they each received from other
users with the number of messages they were sent in
a month.

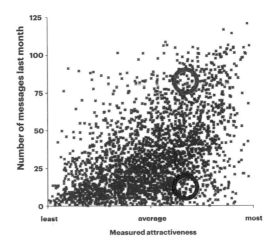

The results are shown in the graph above. Each black cross relates to one user, with the most popular women at the top of the graph, while the most attractive sit along the right-hand side. It might look like a bit of a mess at first glance, with data points all over the place. But the scattered effect of the data says something very interesting: It's not just the better-looking people who receive lots of messages.

But if a high average attractiveness score isn't enough to make you popular, then what is? And what is the difference between the desirable singles at the top of the graph (in blue) and the unpopular people at the bottom (in red), despite the fact that they are considered equally attractive?

The answer, discovered by the OkCupid team, lies in how people view your attractiveness, and is best explained with an example. Imagine scoring two particularly lovely lady cartoon characters: Wilma Flintstone, and Leela from *Futurama*.

I think we can all agree that Wilma is an extremely beautiful woman—nobody could think that she is ugly, but it's fair to say that she's also not exactly Jessica Rabbit, either.

Compare her to how someone like Leela might score. There are some people, myself included, who think that Leela is a seriously sexy lady. But there are others who might be a bit put off by the whole one-eye thing.

I would guess that on average, both women would score roughly the same on a 1 to 5 scale, but the way that single cartoon characters looking for love would score the girls would be quite different. Wilma's scores would all be clustered around 4, but you'd expect a huge spread in the way that people voted for Leela.

Curiously, it's this spread that counts. People like Leela who divide opinion end up being far more popular on internet dating sites than people like Wilma who everyone agrees is "quite cute."

This effect really pops out in the data from the real OkCupid users when you subject it to some statistical sorcery. Using a technique known as regression analysis, the OkCupid team used the data to derive an equation for the number of messages each user can expect, based on how people score them on their attractiveness:

$$Messages = 0.4\,a_1 - 0.5\,a_2 - 0.1\,a_4 + 0.9\,a_5 + k$$

Here a_1 is the number of people who rated you as a 1 out of 5 on attractiveness, a_2 is the number of people who rated you as a 2, and so on. The final value, k, is how active you are on the website. The numbers in front of each part of the equation (or "terms," if we're being proper) come directly from the data and signify how the attractiveness ratings that you are given affect how many messages you can expect to receive.

The +0.9 immediately preceding the a_5 term means that for every one hundred people who rate you as a smoking hot 5 out of 5 you can expect to receive ninety extra messages in a month. So, lucky for you.

It makes sense that getting 5s on attractiveness would equate to receiving more messages. But surpisingly, the +0.4 preceding the a_1 term means that people on OkCupid also receive forty more messages for every one hundred people that score them as 1 out of 5. Yes, you read that right. Having people think you have a face like a dog's dinner means you get more messages.

By contrast, the −0.1 in front of the a_4 term means that users receive ten *fewer* messages for every one hundred people who rate them as a 4 on attractiveness. Having people score you as a 4 out of 5 on attractiveness actually works against you.

To summarize, as long as some people think you're beautiful, you're much better off having some other people think that you're ugly than having everyone think you're just quite cute. People who are unbelievably good-looking and score straight 5 out of 5s will always do well, of course, but the rest of us would do better to divide opinion than to aim for being the cute guy or girl next door.

It seems like quite a counterintuitive result, but maybe what's going on here is that the users sending the messages are also thinking about their own chances: If they think you're beautiful, but suspect that others might be uninterested, there's less competition and therefore an extra incentive to get in touch. But if they think you're beautiful and feel sure that everyone else will, too, they might imagine you'll be getting lots of messages and decide not to bother humiliating themselves.

And here is the really interesting part, because when most people choose their online dating profile pictures, they tend to try and hide the things that make them unattractive. Classic examples include overweight people choosing a really cropped photo, or bald men choosing pictures of themselves wearing hats. But this is the exact opposite of what you should do. When choosing a profile picture, you should play up to whatever makes you different—including the things that some people might not like.

The people who fancy you will still fancy you. And the unimportant people who don't will only play to your advantage.

So be proud of that bald patch, show off that ill-advised tattoo, and get that belly out. Because standing out online just means being yourself. Who would have thought?

5 The Dating Game

Now that you have read chapter 4, let's assume that you're a roaring success on the internet dating scene, with one of the best profile pictures around. But how do you turn your online success into off-line glory? Are there mathematical rules that can help us get what we want out of dating? Of course there are.

But let's park any ideals of embarking on a relationship based in mutual respect and compassion for a moment. Because—understandably—some people have a very clear idea of what they want from a romantic encounter and aren't afraid to go out and get it. This is the motivation of international bestsellers like *The Game* and *The Rules*, which have paved the way for men and women to treat each other as conquests. And both are based on a single idea: how to exploit stereotypes to try and maximize your own reward.

As we've already seen, the mathematics of game theory can be used to beat other suitors. And if you're looking to turn the dating game into a dating war, it is also ideally placed to provide the best strategy in a romantic contest between two opponents.

A warning: Game theory encourages you to exploit the weaknesses of your opponents. When applied to dating, this view comes with a slightly cynical picture of the world.

As a result, the first half of this chapter will show you some of the best tenets of game theory, not the best tenets of human morality. And because they rely on exploiting the supposed differences between men and women, they don't really work for any nontraditional or nonheterosexual couples. Apologies if this doesn't apply to you, but I think it's fair to say you really won't feel left out by the end.

So that I can sleep at night, I've also included a much more sensible and realistic example of how to deal with general dating conundrums, whatever kind of relationship you're in, toward the end of the chapter. But first, let me begin with an illustration of how game theory can be used by men with only one thing on their mind.

Getting what you want out of women

Gentlemen: Your challenge, if you choose to accept it, is to try and persuade women to have sex with you. To assist you on your quest, two mathematicians, Peter Sozou and Robert Seymour, have come up with a strategy you might want to try. They assume you have a range of gifts at your disposal—ones you can offer to sweeten the deal. Think diamond rings or theater tickets. Your task is to choose which gifts to give that are most likely to get you your reward, but without attracting dangerous gold diggers.

Meanwhile, game theory has given you an opponent: the woman, who can decide whether or not to accept the gift. Her task is to try and capture the best man possible, using sex as a bargaining tool and gifts as a reward. Based on the value of the gift, she will try to work out the man's intentions. If she decides that he is likely to stick around, or has sufficiently demonstrated his wealth or attractiveness, she might well sleep with him.

Again, I should add that I don't quite agree with this view of the world (although I am a bit scared that it might be true), but these assumptions do make for a wonderfully neat mathematical problem. The full derivation of the best strategy for the man [1] gets quite heavily into game theory at points and is not for the fainthearted, but the result is a great example of the theory in action. And the best strategy for how to woo the ladies while avoiding gold diggers makes good intuitive sense:

To impress the girl, the man should display showy and extravagant behavior, making purchases that are costly to him, but ones that are ultimately worthless to the female. So, gents: If you want to demonstrate how wealthy you are, put on a big fireworks display or arrive at her house in a Ferrari. If you want to show her how generous you are, leave a big tip at dinner. But don't, whatever you do, buy her jewelry or take her to see her favorite band. She needs to see that the display is expensive to show that you mean business. But it shouldn't be a gift that is valuable to her, or she may string you along without ever intending to have sex with you: classic gold digger behavior.

This theory also works to explain why it's worthwhile for companies to make big, extravagant displays of power—like the marble entrances to Wall Street banks or opulent skyscrapers in Vegas. The bigger a waste of money these gestures appear to be, the more powerful

1. Sozou and Seymour, "Costly but worthless gifts facilitate courtship" (2005).

the consumers and competitors will think the company is. According to the theory, this makes a lot more sense than spending the money buying small gifts for your many customers—to do that risks being exploited by some people who will take the gifts and run without ever having a serious intention of doing business with you.

Speaking as a big fan of diamonds and the White Stripes (hint hint), I'd like to point out that I don't really buy this theory when it's applied to dating. I think it fails to capture something important at the heart of human courtship. Sometimes you're not just out for what you can get; sometimes it's just nice to give nice gifts to people that you like. You know, because of "happiness" and "kindness," etcetera.

Now, girls. I know you are all feeling left out with this example being so focused on what the men should do. But fear not, there are plenty of slightly patronizing applications of game theory out there to help you snare your own rewards, too. Because if men are only after sex, of course we women are *constantly* trying to trick men into marrying us.

Getting what you want out of men

In the age-old game of courtship, men are supposed to be the hunters and women the hunted. But now that I've reached my thirties, there does seem to be a disparity between the number of beautiful, intelligent single women still on the market and the number of handsome eligible bachelors. I'm not the only one to have noticed this, and cries of "where are all the good men" are now heard just as often in London and Shanghai as New York. But this disparity doesn't make sense mathematically. Shouldn't there be the same number of both?

In what became known as the eligible bachelor paradox, Mark Gimein offered an answer to this question using game theory, with a set of assumptions as follows.

Throughout the course of his life, a man will date a range of women. Because of their looks or intelligence or social status, some of these women will be "strong" candidates for settling down with, others, less so. The man will choose to ask one of the women to marry him based on how much he likes her, but also based on how much the woman bids for his affection.

When framed in this way, the game of dating is mathematically equivalent to what happens in a particular kind of auction where bidders submit sealed bids and no one knows the bid of any other participant. The theory starts with two bidders who are both vying for the same lot. One is a strong bidder with access to a lot of cash, the other a weak bidder with a limited budget.

In the case of the bachelor, the man is the lot. The strong bidder is the glamorous, intelligent woman with access to a lot of pizzazz. The weak bidder is less attractive (by whatever measure) and has a more limited charm budget. They are both going for the same man, without knowing how hard the other is trying.

You might think that the strong bidder has the best chance of winning the man, but in real-world auctions it turns out that it's often the bidder in the weaker position who comes away with the prize, a phenomenon which has been the subject of extensive attention within game theory literature.

As with the previous example, the theory[2] gets quite heavy in places, but the insights can go some way toward explaining why there are so many fantastic women in their thirties all competing for a seemingly tiny pool of eligible bachelors.

When a weak bidder comes across a man that she likes, she is likely to pull out all the stops to compete for his attention. A strong bidder, on the other hand, confident that she presents a good match for any man, is less likely to go all-out, knowing that another, better man is probably waiting for her just around the corner.

Seeing disinterest from the more attractive woman, the man will then settle down with the woman who shows him the most attention, taking him out of the dating pool.

This is all fine at the beginning, but as the auction (i.e., life) continues and the lots are won by the weaker bidders, a situation arises with only a few decent men left and a much larger number of beautiful and intelligent women all fishing in the same shrinking pool.

The result is the eligible bachelor paradox, and it comes with a clear, if slightly harsh, take-home message: No matter how hot you are, if your goal is partnership, don't get complacent.

But before we consign ourselves to dying alone and rush out to buy a houseful of cats, it's worth pausing and looking at these examples objectively. As neat an application of game theory as they are mathematically, they have one flawed assumption at their core: that men are trying to trick women into having sex with them and women are desperate for commitment.

2. Güth, Ivanova-Stenzel, and Wolfstetter, "Bidding Behavior in Asymmetric Auctions: An Experimental Study" (2005).

In reality, don't both sexes want both? Crazily enough, I suspect there may even be some women who want sex and some men who want commitment. And thus this particular game theory house of cards comes tumbling down.

Thankfully, there are ways to use game theory that don't require men and women to conform to stereotypes, and in particular, a formulation that can apply to many of the most common dating conundrums for every type of relationship. We'll come to that shortly, but first, let me describe the background theory in a simple example: two people deciding whether or not to cheat on their partner in a relationship.

The game of being faithful

We can set this up with a game between two people in an imaginary relationship: Don (in blue) and Betty (in red). Don and Betty are not moral people; they won't worry about cheating because it's "wrong." Instead, they just want to end up with the highest scores, or "payoffs," from their relationship. These payoffs are determined for each partner by the different strategies they choose to follow, and can be displayed in a table like the one below—what's known in mathematics as a "Payoff Matrix."

	BETTY	
DON	**Faithful**	**Cheat**
Faithful	10 \ 10	20 \ −10
Cheat	−10 \ 20	−5 \ −5

The best outcome for everyone is when Don and Betty manage to maintain a faithful relationship. In that scenario (which is "Pareto optimal"), both parties will get something positive from the relationship. For the purpose of illustration, let's imagine they both gain 10 payoff points in this scenario. Remember, Don and Betty both want to end up with the most points possible from their relationship.

In this game, though, as in reality, there will always be some temptation to cheat on your partner. If Don decides to cheat, he might be able to maintain his relationship with Betty while keeping his bit on the side and increase his own payoff points to 20. Betty, on the other hand, feels the pain of Don's betrayal and her payoff dips to –10 points.

However, the setup of the game is the same for Betty; there is the same incentive for her to cheat, too. But note what happens if both partners decide to give in to temptation and cheat. In that case, everyone loses. Both people end up with –5 payoff points and the relationship breaks down, leaving both parties far worse off than if they had both been faithful.

The numbers here are arbitrary, but the order of the rewards is important. Being the sole cheater results in a higher score than being in a faithful relationship, but it's bad news if someone cheats on you, and bad for everyone if both partners betray each other. Using this setup makes the game of being faithful equivalent to one of the most famous and well-studied problems in game theory: the prisoner's dilemma.

In the prisoner's dilemma, two suspects are being questioned separately about the same crime. They have two choices: to cooperate and stay silent and split the sentence between them, or to defect and rat out their friend. They get to walk away if they talk while their partner stays silent, but they both get a long stretch in jail if they both rat each other out. The reward structure is identical to the game of being faithful: Giving evidence while the other stays silent is better than both staying silent, which in turn beats both cheating. The worst outcome of all is staying silent while your partner rats you out.

This setup does give quite a depressing view of relationships. Cooperation seems difficult to achieve and fragile to maintain. So, if the theory is correct, how is it possible that anyone can have a successful and faithful relationship in a situation this unstable?

The reason is that relationships aren't about one-off decisions. The payoff matrix above doesn't apply to your relationship as a whole. Instead, it's as though both of you are playing this game with each other every day; choosing to cheat or remain faithful on a regular basis. And this difference is key. Repeatedly playing the same game with the same person has a dramatic effect on the way your incentives appear. Suddenly, you're trying to end up with the biggest score over time, rather than in each individual encounter. In the long term you're both better off staying faithful.

What to do when he doesn't call

If you repeatedly cheat on someone, they will no longer trust you. If they believe that you'll always cheat, the only thing they can do for themselves is to cheat back, leaving you both worse off, or alone.

However, if you can find yourself in a situation where you can trust each other to cooperate, you'll both be rewarded at every step of the process. There's little incentive to go for the short-term gain in cheating given how much you stand to lose in the long run.

These ideas were first presented in Robert Axelrod's groundbreaking 1984 book on game theory, *The Evolution of Cooperation*. In it, Axelrod explains how and why cooperation can occur in human and animal societies despite how stark things in Don and Betty's payoff matrix above can at first appear. He also gives a strategy to use whenever you find yourself repeatedly playing this kind of game against the same person.[3]

But Axelrod's strategy doesn't just apply to cheating. It can be used to give you a set of rules to apply in a whole host of dating conundrums. Has your date failed to call you when they promised they would? Has your boyfriend missed your birthday? Should you keep quiet and let things go or flip out over any sign of bad behavior? Axelrod's "tit for tat" strategy gives you the answer.

Despite the name, tit for tat isn't about playground squabbling. It's a strategy that encourages cooperation but punishes exploitation. The mathematical version involves cooperating at first, and then simply copying the previous move of your opponent. They stay cooperative, you stay cooperative. They cheat and defect on you, you cheat and

3. Although Axelrod's tit for tat might not be optimal in all scenarios, it has repeatedly proved extremely successful in several computer tournaments of prisoner's-dilemma strategies. And, as a nice and simple strategy, it does particularly well in the long term—making it ideal to apply to dating.

defect on them. They go back to being nice, you go back to being nice.

Taking the strategy out of the textbook and into the dating world, your rules to follow can be broken down into four simple steps:

1. *Be clear.* Don't play games within the game. Being manipulative or tricky in dating won't work out in the long run. A straightforward strategy will give you the best chance of success.

2. *Be nice.* Start off by being cooperative and continue to be so unless given reason to act otherwise.

3. *Be provokable.* Don't allow yourself to be exploited by bad behavior. If someone treats you badly, you should retaliate with a measured response. But don't go overboard. As soon as a bad deed has been dealt with you should then:

4. *Be forgiving.* Move on from bad behavior quickly and return to being cooperative. You have nothing to gain from continually punishing someone for a single mistake. Going too far with your reaction will only prompt a bad reaction from your partner and find you in a spiral of negativity that can be difficult to recover from. Move on and go back to playing the game together as a team as quickly as possible.

So, to summarize: Don't be a jerk.

Now doesn't that all seem sensible? Don? Betty? At the very least it's much nicer advice to follow than the charmless and, let's face it, chauvinistic advice in books like *The Game*, and it could have a genuinely positive impact on your relationship.

Instead of treating the object of your affection as, well, an object, you could try to follow these simple mathematical rules and act like a sensitive human being. That's why all mathematicians make famously excellent lovers (and dancers). Who knew math could give you such a lovely and moral way to live?

6 The Math of Sex

Once you've found someone you like and wooed them with your charming personality and dashing good looks, things will inevitably lead to the bedroom.

This chapter will not make you better at sex. I thought I should be up front about that, just in case you thought mathematicians kept all the equations for being an amazing lover under their hats. But, if you'll allow me to take a short detour to a more removed perspective, I'd like to share with you some insights into our sexual habits that mathematics can offer. Insights that go far beyond just basic statistics.

Many things can happen when two people have sex for the first time: the start of a new life, the start of a new infection, intense mutual embarrassment, and even, occasionally, pleasure. However, one thing always happens whenever two people have sex: They create a link between themselves in an imaginary network.

These connections can't be taken back, however much one might want to once sobriety returns. They're also two-way (even if the orgasms weren't): Both people will add to their total number of sexual partners whenever a new connection takes place. These clear and well-defined links make the network of sexual contact a particularly interesting case study for scientists and mathematicians.

While we can't see or map the network of connections that we've all inadvertently contributed to by having sex with one another, we can use mathematics to understand its important properties. The network can shed light on the differences between men and women, give us insights into the patterns of human sexual behavior, and even, as we'll reveal later in this chapter, provide a tactic to help stop the spread of sexually transmitted diseases.

Our story begins with a survey conducted by some Swedish scientists in 1996. Through interviews and questionnaires, they collected information on the sexual history of 2,810 randomly selected Swedes from up and down the country, including—crucially—how many people each participant had slept with. As we shall see, the huge number of responses provided the first opportunity for other scientists and mathematicians to study the web of sexual contact, but the original study also came with its own interesting findings.

Magic numbers

Much like several surveys that had gone before, the scientists found that the average number of sexual partners was actually relatively low: around seven for heterosexual women and around thirteen for heterosexual men.

But before we start reinforcing any old-fashioned theories about promiscuous men and chaste women, the eagle-eyed among you might question this discrepancy. And you'd be right to do so. By virtue of the fact that there are roughly the same number of heterosexual men and women in the world and that sex has to occur between two people, the average number of partners for both men and women should be the same. And yet, the difference in male and female averages comes up time and time again in surveys of this kind.

There are a few possible explanations for this difference. Perhaps men are more likely to exaggerate (or "lie," as it's known in the literature). Perhaps men and women have different definitions of what has to take place to add a partner onto their total.

A slightly more persuasive argument revolves around the fact that there may be some women with an unusually high number of sexual partners who are underrepresented in the study. For example, imagine if the next woman they interviewed had slept with 3,000 people. Just this one extra data point would be enough to increase the average number of partners for all females from seven to eight, highlighting again the big issue with using the arithmetic mean to represent the average.

But perhaps more significantly, it appears that the way men and women arrive at their number is different. Women tend to count upward, listing their partners by name: "Well, there was Harry, then Zayn, then Liam . . ." This does tend to give quite accurate results, but if you forget anyone while counting, you are prone to underestimating your true number of partners. Men, on the other hand, are much more likely to approximate: "Say . . . five a year for the last four years." Again, an acceptable method, but it does rather leave you at risk of overestimating. This theory is strengthened when you realize that a surprising number of male answers happen to be divisible by five.

Beyond looking at averages, though, the Swedish study also provided the data for a revolutionary finding.

A formula that unites us

In 1999 Fredrik Liljeros and a team of mathematicians plotted all of the responses from the Swedish survey on a graph and found a startlingly simple underlying pattern. The list of 2,810 responses all lie on a near-perfect curve like that below, showing a clear pattern in the number of partners each participant had admitted to.

Most people had had relatively few sexual partners—which is why the left-hand side of the curve is so high. But there were some responses from people with an

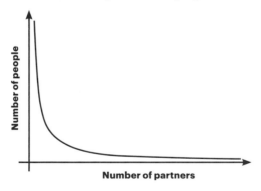

extraordinary number of conquests, which is why the right-hand side of the line on the graph never quite reaches zero. If the Swedish survey is representative of the population at large, the curve suggests that there will always be some chance of finding someone with any number of sexual partners, however large. Granted, there won't be many people in the world with 10,000 or even just 1,000 partners, but the pattern predicts that there will always be some.

All of this can be wrapped up in a single formula that allows you to predict how many people we've all slept with: If you pick a person in the world at random, the chances that they will have had more than x sexual partners is just $x^{-\alpha}$.

The value of α comes directly from the data. To give you an example, the team found the Swedish women had a value of $\alpha = 2.1$. If this number were representative of all of us, the chances of finding someone in the world with more than one hundred partners would be 0.006 percent, suggesting that just over one in 15,800 of us have accomplished that feat. The probability drops the higher the numbers go, but the chances of finding someone with more than 1,000 partners would then be 0.00005 percent, or one in every 2 million people.

Before I completely explode with excitement over the elegance of the mathematics, I think it's worth pausing for a second to appreciate how extraordinary this finding is. For all our free will, and despite the seemingly complicated set of circumstances that lead to our sexual encounters, when you look at the population as a whole there is a startlingly simple formula lying behind everything that we're doing.

This formula means that the number of sexual partners we all have is not completely random. Nor does it follow the normal bell curve–type distribution that is usually associated with things to do with humans, like height or IQ. Instead, the formula suggests that the number of sexual partners follows what's known as a "power-law" distribution.

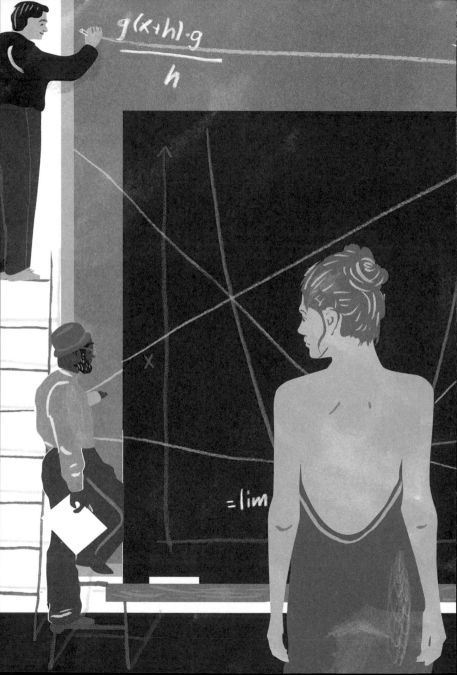

When it comes to height, almost all of us fall within a small window, with most people between five feet and six foot five. There are some outliers, of course, but generally there is little difference between the tallest and shortest people in a typical population. The power-law distribution, on the other hand, allows for a much, much bigger range. If the number of sexual partners followed the same rules as height, finding someone with over a thousand partners would be like meeting a person who was taller than the Eiffel tower.

Partly inspired by this study, scientists and mathematicians have begun to look for and find power-law distributions in a range of unusual places in the last decade. The distribution pattern behind sexual contacts is also found in the way that websites are linked on the internet, the way we form connections on Twitter and Facebook, the way that words sit next to each other in a sentence, even how different ingredients are used in recipes. The simple equation of $x^{-\alpha}$ unites them all.

The reason for all this becomes clearer when we return to the idea of links in a network. It's these connections that are causing the distribution. Power-law distributions are created by links in a network with a very particular shape, known to mathematicians as "scale-free." [1]

An example of what these scale-free networks look like is on the right. Most people have roughly the same number of connections, but there are some—like the darker circle in the middle—who have a huge number of links. These

---

1. These networks are known as scale-free because—unlike normal distributions or Poisson distributions—the underlying power-law doesn't have a typical parameter (like the mean or standard deviation) that defines its scale.

people are known as the "hubs" of the network and are the secret to the similarities between all the seemingly unrelated power-law distributions. Katy Perry, with 57 million followers (as of September 2014), is the biggest hub of the Twitter network, Wikipedia is a hub of the World Wide Web, and the onion is a hub of the recipe ingredient network.

The hubs are created because of a "rich-get-richer" rule in all of these scenarios. The more followers that Katy Perry has, the more likely people are to follow her.

And, if you consider the hub of a sexual contact network, the more successful they are in pursuing sexual conquests, the more likely they are to succeed in persuading more people to sleep with them. They're also the reason that sexually transmitted diseases spread so quickly and are so difficult to control. When a hub doesn't take proper precautions, the hub is the person most vulnerable to contracting a disease, and also the most likely to pass it on. If you can picture a virus spreading through the scale-free network above, you can imagine how crucial the hub is to how things play out.

Hubcaps

But while the hubs are the most critical players in the spread of disease, there's a mathematical trick that allows us to exploit them and the structure of the network when trying to halt the progress of a sexually transmitted virus:

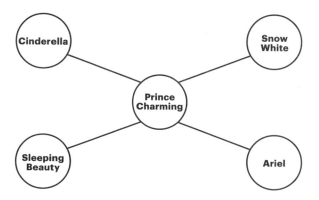

The theory becomes clear if you imagine a simplified network.

Imagine that four beautiful young princesses, Cinderella, Snow White, Ariel, and Sleeping Beauty have each been getting it on with one sexy prince and formed a sexual contact network. There have been no encounters between the ladies on this occasion, unless you count some seriously sketchy Disney fan sites (which I recommend avoiding if you value your childhood memories).

Now imagine there is some kind of nasty sexually transmitted disease going around the group. If vaccinating or educating everyone is too expensive, we might want to prioritize and target our attention only on the hub: the person likely to have the biggest impact.

But without asking everyone how many people they've slept with, we would have no way of seeing the hidden links in the underlying network, and no idea that Prince Charming was our hub.

The task, then, is to try and maximize our chances of finding the hub without knowing the underlying network.

If we picked someone at random to vaccinate in this group, we'd only hit the hub one in five times.

But imagine that instead, we pick someone at random, say the lovely Ariel, and ask her to help us vaccinate someone who she has slept with. Ariel will take us to Prince Charming. Likewise, if we randomly picked Cinderella, and asked her to tell us someone she has slept with, she'd also take us to Prince Charming. So would Sleeping Beauty, and so would Snow White.

Just by adding this one simple step to the algorithm, we increase our chances of finding the hub to four out of five. Much better odds.

The same would be true of much larger networks. Imagine that, without being able to see any of the network or follower statistics of Twitter, we were trying to find Katy Perry—the biggest hub at the time of this writing.

If we picked someone at random from the 500 million people on Twitter, we'd only have a one in 500 million chance of finding Katy.

But, if we picked someone at random and asked them to point us to the most popular person they follow, it would take us to Katy a cool 57 million times. Suddenly the chances of finding Katy soar to around 10 percent, which is pretty impressive given how simple the algorithm is.

This procedure has been used to forecast and slow the

course of epidemics without the need for a difficult and expensive survey of the underlying network. But it also says something impressive, I think, about the simplicity of the vast network that connects us all and how, armed with a mathematical understanding and a basic algorithm, we can gain an important perspective on how sexually transmitted diseases spread.

So next time you add another notch to your bedpost, consider the immense network to which you are contributing. Mathematicians can't help you have better sex, but we do try and cut down the number of STDs you might catch, and what isn't sexy about that?

7 When Should You Settle Down?

When it comes to love, making long-term decisions is a risky business. Sooner or later, most of us decide to leave our carefree bachelor or bachelorette days behind us and settle down. Our wild oats, if we had any, have been sown, and it's time for a lifelong partnership. But how do we know when we've truly found "The One"? As any mathematically minded person will tell you, it's a fine balance between having the patience to wait for the right person and the foresight to cash in before all the good ones are taken. Just ask anyone who has found themselves stung by the eligible bachelor paradox.

If you decided never to settle down, you could sit back at the end of your life and list everyone you ever dated, with the luxury of being able to score each one on how good they could have been as your life partner. I'll admit that such a list would be pretty pointless by then, but if only you could have it earlier, it would make choosing a life partner a fair sight easier.

These potential matches are out there in the world waiting for you to discover them. The list kind of does exist, if only in an imaginary way. But the big question is, how can you select the best person on your imaginary list to settle down with, without knowing any of the information that lies ahead of you?

Let's assume for a moment that the rules of dating are simple: Once you decide to settle down and take yourself out of the dating game, you can't look ahead to see all the partners you could have had on your list, but equally, once you reject someone, you can't go back and change your mind at a later date. That's been my experience, anyway: People seem surprisingly averse to being called up several years after being rejected because no one better came along.

When dating is framed in this way, an area of mathematics called "optimal stopping theory" can offer the best possible strategy in your hunt for The One. And the conclusion is surprisingly sensible:

Spend a bit of time playing the field when you're young, rejecting everyone you meet as serious life-partner material until you've got a feel for the marketplace. Then, once that phase has passed, pick the next person who comes along who's better than everyone you met before.

But optimal stopping theory goes further. Because it turns out that your probability of stopping and settling down with the best person (denoted by P in the equation below) is linked to how many of your potential lovers (n) you reject (r), by a rather elegant formula:

$$P(r) = \frac{r-1}{n} \sum_{i=r}^{n} \frac{1}{i-1}$$

This formula, innocent as it seems, has the power to tell you exactly how many people to reject to give you the best possible chance of finding your perfect partner.

It tells you that if you are destined to date ten people in your lifetime, you have the highest probability of finding

The One when you reject your first four lovers (where you'd find them 39.87 percent of the time). If you are destined to date twenty people, you should reject the first eight (where Mister or Miz Right would be waiting for you 38.42 percent of the time). And, if you are destined to date an infinite number of partners, you should reject the first 37 percent, giving you just over a one in three chance of success.[1]

I know I'm a mathematician and therefore biased, but this result literally blows my mind. If you chose not to follow this strategy and instead opted to settle down with a partner at random from your destiny, you'd only have a $1/n$ chance of finding your true love: just 5 percent if you are fated to date twenty people in your lifetime. But, just by rejecting the first 37 percent of your lovers and following this strategy, you can dramatically change your fortunes, to a whopping 38.42 percent for a destiny with twenty potential lovers.

Okay, before I get too carried away: You may well have spotted some of the flaws in this plan when applying it to dating. Unless you're a member of the English royal family in the 1500s, your potential dates won't be lined up ahead of time, and there's no way you could possibly know how many people will be available to you in your lifetime. And unless you're actually Hugh Hefner, you're probably not going to date an infinite number of people, either.

Thankfully, though, there is a second version of this problem that is much more suited to mere mortals like you and me, and it's got an equally impressive result. Instead

1. As n approaches infinity, the sum may be approximated by an integral with $P(1/e) = 1/e$ for Euler's number e.

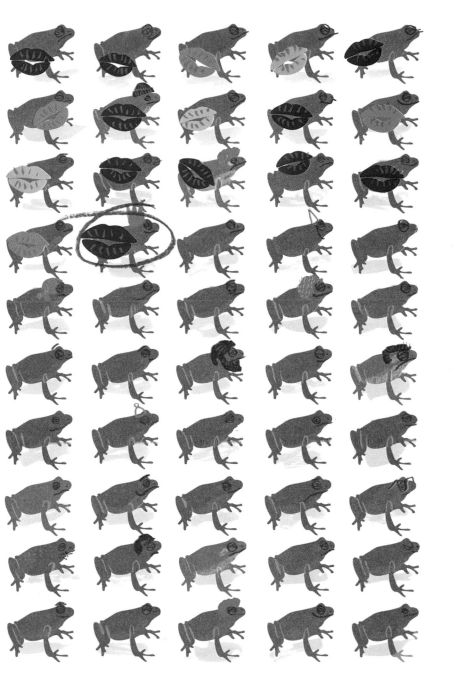

of knowing how many people you'll date, the advanced problem only requires you to know how long you expect your dating life to be. The math in this example is much trickier, [2] though the same simple rule as earlier crops up again—but this time, the 37 percent applies to time rather than people.

Say you start dating when you are fifteen years old and would ideally like to settle down by the time you're forty. In the first 37 percent of your dating window (until just after your twenty-fourth birthday), you should reject everyone; use this time to get a feel for the market and a realistic expectation of what you can expect in a life partner. Once this rejection phase has passed, pick the next person who comes along who is better than everyone who you have met before.

Following this strategy will definitely give you the best possible chance of finding the number one partner on your imaginary list. But, a warning: Even this version of the problem has its flaws.

Imagine that during your 37 percent rejection phase you start dating someone who is wildly charismatic, devastatingly handsome, and a brilliant conversationalist: your perfect partner in every possible way. But, not yet having met everyone, you'd have no way of knowing that they were the best of your list. If you were following the math, you'd have to stick to the rules of the rejection phase and let them go. Unfortunately, once the rejection window had passed and you started looking more seriously for a life

2. I'd like to explain it properly, but it really does get quite complicated. And let's face it, we've all got lives to be getting on with.

partner, no one better would ever come along. According to the rules, you should then continue to reject everyone else for the rest of your life, grow old, and die alone, probably nursing a deep hatred of mathematical formulas.

Likewise, imagine you were really unlucky and everyone you met in your first 37 percent was insufferably dull and boring. Thankfully, you'd be within your rejection phase, and so wouldn't end up spending a lifetime with them. However, now imagine that the next person you dated was still terrible, just marginally less terrible than those before. If you were still following the math, sadly, I'm afraid you'd have to marry them and find yourself trapped in a suboptimal marriage.

Considering all the risks, though, this is still the best possible strategy available for our simplified rules of dating, and I think it still rings true with how many people act in reality. Often, we choose to date a few people first and not think seriously about finding a life partner until we hit our mid- to late twenties. In Europe, women get married at twenty-seven and a half on average, which fits in quite nicely with the theory. I can imagine that men are a little more relaxed about the upper limit of when they would like to be settled, and this certainly seems to be reflected in the average age of marriage for men of thirty and a third across Europe.

Beyond choosing a partner, this strategy also applies to a host of other situations where people are searching for something and want to know the best time to stop looking. Have three months to find somewhere to live? Reject everything in the first month and then pick the next house that comes along that is your favorite so far. Hiring an assistant? Reject the first 37 percent of candidates and then give the job to the next one who you prefer above all others. In fact, the search for an assistant is the most famous formulation of this theory, and the method is often known as the "secretary problem."

Despite its applications, and my many caveats, I may still have oversold this "reject 37 percent" strategy in the context of dating. Because there is still one flaw I haven't yet highlighted. So far, the math assumes you're only interested in finding the very best possible partner available to you. But the situation changes slightly if you're a bit more relaxed about who you end up with. In reality, many of us would prefer a good partner to being alone if The One is unavailable. What if you would be happy with someone who was just within the top 5 percent or 15 percent of your potential partners rather than insisting on all or nothing?

Mathematics can still offer some answers. We can explore the best strategy in each of these scenarios using a trick known to mathematicians as a Monte Carlo simulation. The idea is to set up a sort of mathematical Groundhog Day within a computer program, allowing you to simulate tens of thousands of different lifetimes, each with randomly appearing partners of random levels of compatibility. The program, acting as if on a virtual quest for love, can experiment with what happens in each lifetime if they use a different rejection phase from the 37 percent outlined above. At the end of each simulated lifetime and with the benefit of hindsight, the program looks back at all the partners it could have had and works out if the strategy has been successful.

If you repeat this process for every possible rejection phase, for each of the three criteria of success (best partner only, someone in the top 5 percent, someone in the top 15 percent), you end up with a graph that looks like this:

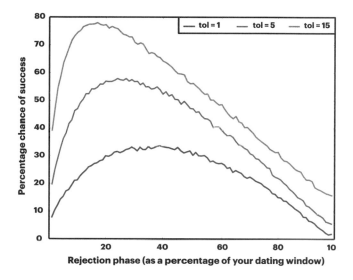

The red line is our original problem. Here, the highest possible chance of success comes with a rejection window of 37 percent as the math predicted, also giving you a 37 percent chance of settling down with the perfect partner.

But if the criteria are relaxed a bit and you'd be happy to settle down with someone in the top 5 percent of everyone you will date in your lifetime, it's the yellow line you need. Here, you have your best chance of success if you reject partners who appear in the first 22 percent of your dating window and pick the next person that comes along who's better than anyone you've met before. Follow this strategy and you'll settle with someone within the top 5 percent of your potential partners an impressive 57 percent of the time.

And if you're much less fussy and would be happy with anyone in the top 15 percent, you only need to spend a mere 19 percent of your dating life getting a feel for what's out there, as in the blue line. Use this strategy and you can expect a whopping 78 percent chance of success—much less risky than the traditional all-or-nothing version of this problem.

These ideas still aren't perfect. Life partners aren't like houses or assistants, which are yours if you can afford them. But nonetheless, I think this neat and simple problem offers some nice insight into the real-world scenario, even if you can't totally take it at face value. After all, that's what mathematics is about—abstracting from the real world to help uncover some of the hidden patterns and relationships which otherwise would be shrouded in messy things like "emotions."

8 How to Optimize Your Wedding

Now that we know how to aim for the perfect partner, let's hope that we'll all have the chance to settle down into a happy and successful lifelong partnership. But for those who decide that marriage is for them, there's another hurdle to overcome before the happily ever after can begin. And once the excitement of the engagement is over, you're left with the ugly reality of having to plan the actual wedding.

No blushing young romancer dreams she'll one day become Bridezilla; no prospective groom imagines he'll be the one throwing a tantrum over color-coordinated table linens. But with so many competing considerations—the in-laws, the budget, the dresses, the venue, the bridesmaids—there's a little bit of crazy in all of us. (Believe me, I speak from bitter experience.)

But before you lose your mind staring at calligraphy fonts and organza chair bows, I want to try and show you how math can help to make things go a little bit more smoothly on the big day.

Mathematical invitations
One of the first things to deal with is the dreaded guest list, something that always turns out to be a harder task than it first appears. Ideally you'd invite everyone you've ever met,

but the realities of budget and venue size quickly find you making tricky decisions between people with a seemingly equal right to be invited.

The people you do invite often come with partners and families who, depending on the severity of your no-ring-no-bring policy, might take precedence over the lonely and single of the invite pool.

Even once you've made these decisions, not everyone who you do end up inviting will turn up on the day. Ensuring you end up with the perfect number of guests is a tricky balancing act. Too few and you've turned away important people who could have otherwise joined the party, too many and you'll find yourself over budget and under pressure for space.

Most people deal with this problem by sending invitations out in phases, adjusting the numbers as the RSVPs roll in. But is this a safe approach to take in an age when people think that their morning's breakfast warrants a Facebook status update? An invitation to your wedding will almost certainly make it into the public realm, alerting second-tier friends and family that they've been left out of the first round of invites.

An alternative strategy could be to underinvite, or not to book the venue until you know the numbers. Or, you could use the strategy employed by most people in this situation: total blind guesswork.

But there is a way to use math to give you a sensible starting point before the arguments with the in-laws begin.

The process starts with a list of all the potential invitees, grouped as couples or families and ordered by how much you want them to be there on your big day. This might already seem like an arduous task, but if you don't know yourself how much you like your friends, there's not much that math can do to help you.

Put this list in a spreadsheet, with the name of the group in the first column, and the number of people the group represents in the next.

The next step is to decide how likely each group of people is to actually show up if you invite them. Think about how far away people live. What else is going on in their lives? Do they secretly hate you? You get the idea.

Think in terms of a percentage but write it down as a decimal. For example, your close friend from home and her boyfriend might be 95 percent likely to attend, so they would collectively get a score of 0.95. This makes up the third column of your spreadsheet.

Multiplying the second column—the number of people in each group—by their score in the third column gives you a fourth column containing the "expected" number of people who will RSVP with a yes.

As you go down the list, from the VIPs to the also-rans, keep a running total of the expected number of guests in the fifth column. The simplest method is then to set the cutoff once the running total of expected guests exceeds the capacity of the venue. By doing this you will, on average, invite the right number of people to your wedding.

The table on the following page gives you an example of how the bottom end of this spreadsheet might look. With a venue that can host one hundred guests, you should invite everyone up to Gordon and family. You'll send out invitations to more than one hundred guests, but on average you can expect only one hundred to arrive. Unfortunately, this

Names	NUMBER OF GUESTS X	PROBABILITY THEY'LL RSVP WITH A YES P(X)	EXPECTED ATTENDANCE E(X)	RUNNING TOTAL
John and family	4	0.95	3.8	94.8
Tony and Cherie	2	0.20	0.4	95.2
Gordon and family	5	1.00	5.0	100.2
David and Sam	2	0.80	1.6	101.8

time David and Sam don't make the cut, which is probably for the best.

The sharp-eyed reader will have spotted a problem with the idea of getting the right number on average. Because we're dealing with probabilities, your final attendee list is just as likely to be over capacity as under. Being underattended gives you the chance to include all those plus-ones and forgotten guests at the last minute, but being overattended might spell disaster on the day.

The advanced version of this idea, then, is to calculate the chances of the worst-case scenario and adjust where you draw your cutoff to leave you with only a small chance of disastrous overattendance.

But how do you work out the probability of disaster?

Imagine you needed to invite 150 guests to give yourself an expected attendance of one hundred people on the day. In reality, you could find yourself with anything from zero to 150 people accepting the invitation, but the chances of either of these extreme situations coming to pass would be very low.

Indeed, it's quite simple to calculate the chances of *everyone* attending—you simply multiply all the probabilities together from the third column in your spreadsheet. For example, the probability of John, Tony, and Gordon and their plus-ones all attending is $0.95 \times 0.2 \times 1.0 = 0.19$, or 19 percent.

In theory, you could calculate the probability of any final number of guests simply by going through the chances of every combination of yeses and nos.[1]

If you were to plot all of these probabilities in a graph, it might look something like the one below. The final guest totals nearer the middle will have a much higher probability, and on average you can expect one hundred guests to arrive.

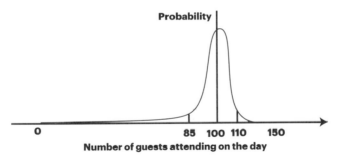

Number of guests attending on the day

Now deciding on a sensible, safe buffer zone is an easier challenge. If you invite 150 people, you can be fairly confident that the actual number of guests will most likely be somewhere in the peak of the curve—between 85 and 110, in this example.

1. Although using a Monte Carlo computer simulation would be much more sensible. Monte Carlo methods offer a way of sampling without having to check every possible combination.

You could then redo this graph to see how the curve, and hence the upper and lower limits, would change if you only invited 130 people instead of 150, or 120 instead of 130, and so on, repeating until you found a buffer zone with a likely best and worst case scenario you were comfortable with. This method has real-world applications. In 2013, two mathematically minded wedding planners, Damjan Vukcevic and Joan Ko used this exact technique to plan their guest list. In their case, they split their guests into four categories and estimated a probability for each category. They sent out 139 invitations and their model told them to expect 106 guests, with 95 percent confidence that between 102 and 113 people would attend.

As it turned out, 105 people came to the wedding, although only 97 of these were on the original invitation list. Damjan and Joan managed to get the correct number of guests, despite making two errors that balanced each other out: They overestimated the probability that friends based locally would attend, but they also underestimated the number of people who would expect to be added to the guest list at the last minute.

As we saw in chapter 1, this balancing of errors is a recurring theme in estimation problems and one of the reasons why assigning a probability to each group on your guest list is a good idea. You'll no doubt be overly optimistic about some guests, but underestimate on others. You might get a bit unlucky, but on average things will work out okay in the end. It's not possible to come up with a method that is *completely* without risk. But it does provide a useful starting point from which to tweak and adjust before finally devising your invitee list.

Mathematical table planning

Unfortunately, though, there are other errors when it comes to weddings that aren't so easy to forget. And aside from a terrible best man's speech or ill-fitting bridal gown, one of the hardest errors to forgive is sitting two enemies together over dinner.

The table plan forms a crucial part of any wedding. Your guests' enjoyment of the day rests largely on your decision of where to seat them. Get it right and you'll successfully bring together friends of both bride and groom. Get it wrong and it'll be difficult to stop the flow of resentment seeping through the room, or the inevitable brawl going on outside.

You need to seat couples and families together, friends on the same table where possible, and enemies apart at all costs.

But this is where the mathematics of optimization comes in. Allocation problems—much like this one—crop up everywhere. Every time you hear claims of something being the best, the cheapest, or the most efficient, there's usually an optimization algorithm at work. And these algorithms, used by everyone from governments to hedge funds to Walmart, can be used to take real measures to ensure that there are no seating-related fights at your wedding.

To decide on the best table plan, it's important to first define what you mean by "best." It could mean maximizing the happiness of your VIPs, or maximizing the average happiness of all guests. You could even aim to minimize the happiness of the guests who you secretly hate but had to invite because you were being polite.

Any of these would be possible to achieve (although I don't recommend the last one), but let's assume that you're aiming for the overall happiness of the room to be as high as possible.

To get this right we need to define what we mean by "happiness."

A simple way to do this is to set up a chart comparing every guest with every other and assigning a score between them describing how they would feel sitting next to each other. Use a positive score whenever two people know each other or would be happy to sit together. The bigger the score, the more important it is that people should be at the same table.

People who don't know each other score zero, and people who should be kept apart should be given a negative score. A large negative score can be used wherever people should be kept apart at all costs.

We can try this with one particularly fraught example of a wedding with just two tables, sadly not taken from real life but instead picking completely at random from a big bag of names.

	LUKE	BRUCE	DALMATIAN PUPPY	DARTH	THE JOKER	CRUELLA
Luke	—	20	60	-20	-5	0
Bruce	20	—	40	-10	-30	-5
Dalmatian puppy	60	40	—	-10	-30	-40
Darth	-20	-10	-10	—	30	15
The Joker	-5	-30	-30	30	—	20
Cruella	0	-5	-40	15	20	—

The answer in this case is obvious: Sit Luke, Bruce, and the puppy at one table and the party poopers—Darth, the Joker, and Cruella—at the other.

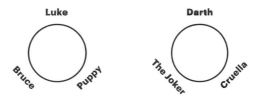

Reading down Luke's column, we see that he would score 20 "happiness points" for sitting next to Bruce and 60 for sitting with the puppy, giving him a total of 80 points.

Similarly, Bruce will get a total of 60 happiness points, and the puppy will have a lovely time with her two new friends and get a score of 100.

On the grumpy table, Darth gets 45 happiness points, the Joker 50, and Cruella a total of 35. At least they're happy being miserable together. Adding up the scores for all the guests, the table plan scores a total of 370 points. Not a bad start for a party.

But switch just one pair of guests and catastrophe strikes. If the dalmatian puppy swaps places with Darth (so that Luke, Bruce, and Darth are on one table and the puppy, the Joker, and Cruella are on the other), the scores will crash to –120.

This example is quite simple, and the ideal table plan is obvious from the outset, but this technique of calculating scores between people does provide a way to methodically address the much larger, and more realistic, table plans of bigger weddings.

The basic method is the same, and in theory, you could check every possible combination of seating arrangement by hand. Problem solved, then.

Oh. Except that even for a small wedding of seventeen people and two tables of ten there are 131,702 different ways to seat your guests.

A computer program that could check one arrangement a second would take over two weeks to run through every possible combination. The decades it would take to do it with a pad and paper might put off your partner. This computation time gets bigger and bigger the more guests you have. A wedding of one hundred guests and ten tables will have 65 trillion trillion trillion trillion trillion trillion trillion different possible table plans. Good luck checking through those before the big day.

This is where the tricks of mathematical optimization really come into their own. There are a host of clever mathematical methods to let you skip across huge chunks of bad combinations without checking them.[2] This means that rather than calculating the overall score for every possible

---

2. Examples include the simulated annealing algorithm and the Nelder-Mead simplex procedure, which both provide efficient ways to search for optimal solutions.

table plan, you can quickly and efficiently search through the combinations and find the best one without checking them all.

Meghan Bellows and J. D. Peterson used this strategy to draw up their own wedding table plan in 2012. They started off by giving a happiness score to each of their 107 guests.

Given the size of the problem, they decided against the pad and paper. They did what any self-respecting wedding planners would do and used General Algebraic Modeling System (GAMS) software with a CPLEX solver to do the work for them.[3] Within thirty-six hours they had a table plan.

If your knowledge of numerical-optimization computer programming techniques isn't quite up to scratch, you should be able to work out one or two of the more difficult tables by hand. Otherwise, simply enlist the aid of your friendly neighborhood mathematician. I find they're usually more than happy to help.

It's impossible to say that the result will always be perfect. The math can only do as well as the numbers you give it. But it should give you a good starting point before you hand your seating plan over to the in-laws, and then the real arguments can begin.

---

3. The CPLEX solver employs an algorithm from linear programming and uses the happiness scores to create a feasible region in solution space. It skips over anything on the inside of this convex polytope by assuming the optimal result will lie on the surface of the simplex.

9 How to Live Happily Ever After

We all love a good wedding. But as depressing as it is to contemplate on someone's big day, it's a sad fact of modern life that many marriages won't make it over the long term.

Although most people manage to remain pretty optimistic about their own chances of success, the non-fairytale reality of just how hard relationships can be is difficult to avoid forever. Regardless of whether you choose to get married or not, wouldn't it be nice to have some understanding of how best to behave in a long-term relationship in order to give you the best chance of staying happy? Perhaps some tools to deal with conflict efficiently and avoid disastrous spirals of negativity? Or a strategy that lets you retain your individuality while keeping you both on the same team?

To address these questions, I wanted to show you one of my favorite applications of mathematics to the story of love—one that is firmly anchored in reality. It comes from a wonderfully impressive collaboration between mathematicians and psychologists and ends with a powerful message about how the mathematical patterns in our real-world relationships tell us to treat one another on the path to living happily ever after.

The mathematics of marriage

Every relationship will have conflict, but most psychologists now agree that the way couples argue can differ substantially, and can work as a useful predictor of longer-term happiness within a couple.

In relationships where both partners consider themselves as happy, bad behavior is dismissed as unusual: "He's under a lot of stress at the moment," or "No wonder she's grumpy, she hasn't had a lot of sleep lately." Couples in this enviable state will have a deep-seated positive view of their partner, which is only reinforced by any positive behavior: "These flowers are lovely. He's always so nice to me," or "She's just such a nice person, no wonder she did that."

In negative relationships, however, the situation is reversed. Bad behavior is considered the norm: "He's always like that," or "Yet again. She's just showing how selfish she is." Instead, it's the positive behavior that is considered unusual: "He's only showing off because he got a pay raise at work. It won't last," or "Typical. She's doing this because she wants something."

Beyond these qualitative insights, though, one team of academics, led by psychologist John Gottman, has come up with a way to assign scores[1] to how positive or negative a couple can be to one another.

Over a number of decades, Gottman and his team observed hundreds of couples in conversation and measured everything they could think of, from their facial expressions to their heart rates, their skin conductivity,

---

1. Known as the Specific Affect Coding System, or "SPAFF" for short.

and their blood pressure, as well as what each person actually said.

Low-risk couples scored many more positive than negative points on Gottman's scale, while couples that were struggling in their relationship often found themselves spiraling downward into negativity.

Although few of us will have portable skin-conductivity testing kits at home, you can use a simpler version of their technique to look at your own relationship.[2]

Set up a camera and film yourselves talking about a particularly contentious subject for around fifteen minutes. Once you've finished (and any anger has subsided), play back the tape and score yourselves as below for everything you've said that fits into one of the following affect categories:

CATEGORY	SCORE	CATEGORY	SCORE
Joy	+4	Contempt	-4
Humor	+4	Disgust	-3
Affection	+4	Defensiveness	-2
Validation	+4	Belligerence	-2
Interest	+2	Stonewalling	-2
		Domineering	-1
		Anger	-1
		Whining	-1
Neutral	0	Sadness	-1

While resisting the temptation to squabble over your scores, try and see if there are any patterns that

2. A full overview of the scoring system can be found in Coan and Gottman, "The Specific Affect Coding System (SPAFF)" (1995).

emerge. Was there something you said that set off a chain reaction of negativity? Could you be more open to understanding your partner's point of view? I'm certainly not a psychologist, but I think there is something positive to be gained from looking objectively at your own behavior through numbers, and trying to see if there was anything you could have done to promote a more positive discussion.

The more sophisticated scoring system constructed by the academics (an extension of the table to the left) allowed Gottman and his team to predict divorce among couples with up to 90 percent accuracy after observing them in conversation. But it wasn't until they teamed up with mathematician James Murray that they really began to understand how these crucial spirals of negativity are formed and how they develop.

Although Murray's mathematical models are framed in terms of a husband and a wife, they aren't based on any gender stereotypes and could apply equally well to a long-term and/or gay or lesbian relationship. They are a wonderfully elegant example of how mathematics can be applied to patterns in human behavior and can be summarized neatly by the following two equations

$$W_{t+1} = w + r_w W_t + I_{HM}(H_t)$$

$$H_{t+1} = h + r_H H_t + I_{HM}(W_t)$$

These equations might look like gibberish at first, but they're actually describing a very simple set of rules for predicting how positive or negative we can expect the husband and wife to be in the next turn of their conversation.

If we take the top line, the wife's equation, we can break down how these rules play out. The left-hand side of the equation is simply how positive or negative the wife will be in the next thing that she says. Her reaction will depend on her mood in general (w), her mood when she's with her husband ($r_w W_t$), and, crucially, the influence that her husband's actions will have on her (I_{HM}). The H_t in parentheses at the end of the equation is mathematical shorthand for saying that this influence depends on what the husband has just done.

The equations for the husband follow the same pattern: $h, r_H H_t$, and I_{HM} are his mood when he's on his own, his mood when he's with his wife, and the influence his wife has on his next reaction, respectively.

It's worth pausing for a moment to mention that these exact equations have also been shown to successfully describe what happens between two countries during an arms race. So, an arguing couple spiraling into negativity and teetering on the brink of divorce is actually mathematically equivalent to the beginning of a nuclear war.

But that doesn't mean that they've just been shoehorned into a new application without purpose. As they've been shown to accurately capture what happens in both scenarios, the analogy only means that insights found in studying international conflict can give fresh meaning to our understanding of marriages, and vice versa. The connection serves to strengthen the mathematics rather than weaken its meaning.

As in the buildup to nuclear war, the most important

thing in Gottman and Murray's marriage equations is the influence term: the effect that the husband and wife have on each other.

As Gottman and Murray were the first people to apply a mathematical model to marital conflict, they were free to choose how this influence term would look, and decided that the following version fit well with everything that had been observed in real-life couples.

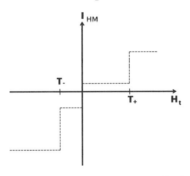

If we take the influence that a husband (H_t) has on his wife (I_{HM}) as a guide, the graph above shows the mathematical model chosen by the team.

Wherever the dotted line is high on the I_{HM} scale, it means that the husband is having a positive impact on his wife. Likewise, wherever the dotted line dips below zero on the I_{HM} scale, the wife is more likely to be negative in her next turn in the conversation.

Imagine that the husband does something that is a little bit positive: He could agree with her last point, or inject a little humor into their conversation. This action will have a small positive impact on the wife and make her more likely to respond with something positive, too.

This happens until a point, $T+$, where the husband does something really nice, like telling her he loves her or agreeing to go with her to that new play she's been wanting to see.[3] Anything more positive than $T+$ will have a big impact on the wife and is much more likely to see the couple draw themselves into a nice, stable conversation with lots of positive reinforcement.

At the other end of the spectrum, if the husband is a little bit negative—like interrupting her while she is speaking—he will have a fixed and negative impact on his partner. It's worth noting that the magnitude of this negative influence is bigger than the equivalent positive jump if he's just a tiny bit positive. Gottman and his team deliberately built in this asymmetry after observing it in couples in their study.

At some point $T-$, though, known as the "negativity threshold," the husband is sufficiently annoying to cause his wife to lose her cool completely and respond very negatively to her husband. This threshold turns out to be quite important in understanding the spirals of negativity that occur in couples.

Now, I always thought that good relationships were about compromise and understanding, and so would have guessed that it was best to aim for a really high negativity threshold. A relationship where you give your partner room to be themselves and only bring up an issue if it becomes a really big deal.

But actually, the team found that the exact opposite was true.

---

3. My husband should take note of this point in particular.

The most successful relationships are the ones with a really low negativity threshold.[4] In those relationships, couples allow each other to complain, and work together to constantly repair the tiny issues between them. In such a case, couples don't bottle up their feelings, and little things don't end up being blown completely out of proportion.

This isn't the end of the story, however.[5] Living happily ever after isn't just about being comfortable complaining. For a start, it's worth adding that the language you use in your conversation should still be open and understanding, and there's always more room to respect your partner as an individual, rather than allowing yourself to feel like the victim of their behavior. But I for one like the idea that mathematics leaves us with a positive message for our relationships, reinforcing the age-old wisdom that you really shouldn't let the sun go down on your anger.

4. Based on a study of newlyweds in Seattle between 1989 and 1992. Along with uninfluenced parameters of both parties, high negativity thresholds were found to be a significant indicator of probable divorce.

5. Readers interested in a comprehensive overview of the academic literature on marriage could do a lot worse than reading the fascinating and brilliantly written book *The Mathematics of Marriage* (Basic Books, 2005), by Gottman, Murray, Swanson, Tyson, and Swanson.

In many respects, this book is simply an homage to the many mathematicians who have dedicated their time trying to tease out and capture the elusive essence of love. Sometimes their pursuits erred more on the playful side of the subject. Sometimes their work can offer sound and sensible advice that applies to us all. From Peter Backus's equation to calculate your chances of finding love to John Gottman and James Murray's mathematical advice for couples, I find each one as neat and elegant as the last.

As different as the efforts to understand love might be, they are all united by a single fact: They all exist only as models of reality. And in the words of the statistician George E. P. Box, "All models are wrong, but some are useful."

It would be easy for some to dismiss the examples in this book as superficial and frivolous, to be cynical about how far they can really apply to love. But I think to do so would be to miss the real insights that they offer. Because, despite their limitations, I think they all combine to tell us something important about mathematics.

Mathematics is about abstracting away from reality, not about replicating it. And it offers real value in the process. By allowing yourself to view the world from an abstract perspective, you create a language that is uniquely able to capture and describe the patterns and mechanisms that would otherwise remain hidden. And, as any scientist or engineer of the past 200 years will tell you, understanding these patterns is the first step toward being able to exploit them.

By being able to describe the behavior of electricity and magnetism, mathematics formed the basis for our modern technological revolution. By providing a platform for rigorous hypothesis-testing and dealing with evidence, mathematics played its role in the modern transformation of medicine. And, as in my own research, mathematics is now being used to study the patterns of human behavior, allowing us to view everything from terrorism to city life from a fresh and insightful perspective.

But just as the best applied mathematicians know the power of their subject, they also know its limitations. They understand the importance of what happens beyond the equations, and they respect the value of other perspectives.

With the financial crash of 2008, we saw the worst of what happens when people misunderstand the weaknesses of mathematical models, when people blindly follow the equations without thought of the warnings and caveats laid out by the mathematicians. To me, these failings reflect a false impression of mathematics that's as grave an error as a mistrust of it all together.

But beyond its limitations, for me, mathematics has a beauty that encapsulates the realistic, the idiosyncratic, and the abstract. And I'll never get tired of finding more hidden patterns and counterintuitive results in the real world, regardless of the assumptions it took to get there.

FURTHER READING

CHAPTER 1: WHAT ARE THE CHANCES OF FINDING LOVE?

Backus, Peter. "Why I Don't Have a Girlfriend." Warwick Economics Summit, 2010.

Drake, Frank. "The Drake Equation" (1961): http://www.activemind.com /Mysterious/Topics/SETI/drake _equation.html.

CHAPTER 2: HOW IMPORTANT IS BEAUTY?

Ariely, Dan. *Predictably Irrational: The Hidden Forces That Shape Our Decisions*. New York: HarperCollins, 2008.

Devlin, Keith. "The Myth That Will Not Go Away." The Mathematical Association of America, 2007.

Johnston, Victor S. "Mate Choice Decisions: The Role of Facial Beauty." *Trends in Cognitive Sciences,* 2006.

Perrett, David. *In Your Face: The New Science of Human Attraction*. London: Palgrave Macmillan, 2010.

Perrett, David I., D. Michael Burt, Ian S. Penton-Voak, Kieran J. Lee, Duncan A. Rowland, and Rachel Edwards. "Symmetry and Human Facial Attractiveness." *Evolution and Human Behavior,* 1999.

Thornhill, Randy, and Steven W. Gangestad. "Facial Attractiveness." *Trends in Cognitive Sciences,* 1999.

CHAPTER 3: HOW TO MAXIMIZE A NIGHT ON THE TOWN

Gale, David, and Lloyd Shapley. "College Admissions and the Stability of Marriage." *The American Mathematical Monthly* 69 (1), 1962.

Huang, Chien-chung. "Cheating by Men in the Gale-Shapley Stable Matching Algorithm." *Algorithms– ESA,* 2006.

McVitie, D. G., and L. B. Wilson. (1971). "The Stable Marriage Problem." *Communications of the ACM* 14 (7), 1971.

Roth, Alvin E., and Marlinda A. Oliviera Sotomayor. *Two-Sided Matching: A Study In Game-Theoretic Modeling and Analysis*. Cambridge: Cambridge University Press, 1992.

CHAPTER 4: ONLINE DATING

Statistics from: http://www.statistic brain.com/online-dating-statistics/.

Ireland, Molly E., Richard B. Slatcher, Paul W. Eastwick, Lauren E. Scissors, Eli J. Finkel, and James W. Pennebaker. "Language Style Matching Predicts Relationship Initiation and Stability." *Psychological Science* 22 (1), 2011.

Rudder, Christian. "Inside OKCupid: The Math of Online Dating" (2013): http://www.youtube.com /watch?v=m9PiPlRuy6E.

———. We experiment on human beings!" (2014): http://blog.okcupid.com/index.php/we-experiment-on-human-beings/.

CHAPTER 5: THE DATING GAME

Axelrod, Robert M. *The Evolution of Cooperation (Revised Edition).* New York: Basic Books, 2009.

Güth, Werner, Radosveta Ivanova-Stenzel, and Elmar Wolfstetter. "Bidding Behavior in Asymmetric Auctions: An Experimental Study." *European Economic Review,* 49 (7), 2005.

Sozou, Peter D., and Robert M. Seymour. "Costly but Worthless Gifts Facilitate Courtship." *Proceedings of the Royal Society B: Biological Sciences* 272(1575), 2005.

CHAPTER 6: THE MATH OF SEX

Bearman, Peter S., James Moody, and Katherine Stovel. "Chains of Affection: The Structure of Adolescent Romantic and Sexual Networks." *American Journal of Sociology* 110 (1), 2004.

Newman, M.E.J. "Spread of Epidemic Disease on Networks." *Physical Review E* 66 (1), 2002.

Liljeros, Frederik, Christofer R. Edling, Luis A. Nunes Amaral, H. Eugene Stanley, and Yvonne Åberg. "The Web of Human Sexual Contacts." *Nature* 411 (6840), 2001.

Pastor-Satorras, Romaualdo, and Alessandro Vespignani. "Epidemic Spreading in Scale-Free Networks." *Physical Review Letters* 86 (14), 2001.

CHAPTER 7: WHEN SHOULD YOU SETTLE DOWN?

Ferguson, Thomas S. (1989). "Who Solved the Secretary Problem?" *Statistical Science* 4 (3), 1989.

Todd, Peter M. "Searching for the next best mate," in *Simulating Social Phenomena,* edited by Rosaria Conte, Rainer Hegselmann, Pietro Terna, 419–36. Berlin: Springer Berlin Heidelberg, 1997.

CHAPTER 8: HOW TO OPTIMIZE YOUR WEDDING

Alexander, Ruth. "A Statistically Modeled Wedding" (2014): http://www.bbc.co.uk/news/magazine-25980076.

Bellows, Meghan L., and J. D. Luc Peterson. "Finding an Optimal Seating Chart." *Annals of Improbable Research,* 2012.

CHAPTER 9: HOW TO LIVE HAPPILY EVER AFTER

Gottman, John M., James D. Murray, Catherine C. Swanson, Rebecca Tyson, and Kristen R. Swanson. *The Mathematics of Marriage: Dynamic Nonlinear Models.* Cambridge, Mass.: Basic Books, 2005.

AUTHOR THANKS

This book isn't exactly *War and Peace*, but it has still required help and support from a number of wonderful people. I owe a huge debt of thanks to James Fulker, Lis Adlington, and Rob Levy—all of whom helped me out of several holes along the way. Equally, Michelle Quint and the TED team deserve a medal for their patience and support throughout this process. My Marge and Parge and sisters Tracy and Natalie are due a good deal of credit, too, not just for this book, but for generally being brilliant people. Huge thanks to Anna Gregson, Peter Baudains, and Thomas Evans—I'm incredibly grateful for your useful comments and ongoing enthusiasm. Andy Hudson-Smith also has my heartfelt gratitude for being so supportive of this and my various other crackpot projects. Thanks to Geoff Dahl, who didn't really help that much with this book, but I just like him as a human being, and to Adam Dennett and Emma Welsh for making me comedy cakes when I needed them the most. And last but not least, thanks to Phil and Miss McGee—I'm very lucky indeed to have you two on my team.

ABOUT THE AUTHOR

DR. HANNAH FRY is a mathematician at the UCL Centre for Advanced Spatial Analysis. In her day job she uses mathematical models to study patterns in human behavior, from riots and terrorism to trade and shopping.

Alongside her academic position, she is currently a UCL public engagement fellow, taking the joy of math into theaters, pubs, and schools. She also copresents the BBC Worldwide YouTube channel and regularly appears on TV and radio in the UK.

Hannah lives in London with her husband, Phil, who—luckily—came along at exactly 38 percent. She has several leftover Python codes from her wedding planning, which can be distributed upon request.

You can find her on Twitter: @fryrsquared

WATCH HANNAH FRY'S TED TALK

Hannah Fry's TED Talk, available for free at TED.com, is the companion to *The Mathematics of Love.*

JAMES DUNCAN DAVIDSON/TED

RELATED TALKS ON TED.COM

Helen Fisher
Why we love, why we cheat

Anthropologist Helen Fisher takes on a tricky topic—love—and explains its evolution, its biochemical foundations, and its social importance. She closes with a warning about the potential disaster inherent in antidepressant abuse.

Esther Perel
The secret to desire in a long-term relationship

In long-term relationships, we often expect our beloved to be both best friend and erotic partner. But as Esther Perel argues, good and committed sex draws on two conflicting needs: our need for security and our need for surprise.

Adam Spencer
Why I fell in love with monster prime numbers

They're millions of digits long, and it takes an army of mathematicians and machines to hunt them down—what's not to love about monster primes? Adam Spencer, comedian and lifelong math geek, shares his passion for these odd numbers, and for the mysterious magic of math.

Yann Dall'Aglio
Love—you're doing it wrong

In this delightful talk, philosopher Yann Dall'Aglio explores the universal search for tenderness and connection in a world that's ever more focused on the individual. As it turns out, it's easier than you think. A wise and witty reflection on the state of love in the modern age.

ABOUT TED

TED is a nonprofit devoted to spreading ideas, usually in the form of short, powerful talks (18 minutes or less) but also through books, animation, radio programs, and events. TED began in 1984 as a conference where Technology, Entertainment and Design converged, and today covers almost every topic—from science to business to global issues—in more than 100 languages.

TED is a global community, welcoming people from every discipline and culture who seek a deeper understanding of the world. We believe passionately in the power of ideas to change attitudes, lives and, ultimately, our future. On TED.com, we're building a clearinghouse of free knowledge from the world's most inspired thinkers—and a community of curious souls to engage with ideas and each other. Our annual flagship conference convenes thought leaders from all fields to exchange ideas. Our TEDx program allows communities worldwide to host their own independent, local events, all year long. And our Open Translation Project ensures these ideas can move across borders.

In fact, everything we do—from the TED Radio Hour to the projects sparked by the TED Prize, from TEDx events to the TED-Ed lesson series—is driven by this goal: How can we best spread great ideas?

TED is owned by a nonprofit, nonpartisan foundation.

ABOUT TED BOOKS

TED Books are small books about big ideas. They're short enough to read in a single sitting, but long enough to delve deep into a topic. The wide-ranging series covers everything from architecture to business, space travel to love, and is perfect for anyone with a curious mind and an expansive love of learning.

Each TED Book is paired with a related TED Talk, available online at TED.com. The books pick up where the talks leave off. An 18-minute speech can plant a seed or spark the imagination, but many talks create a need to go deeper, to learn more, to tell a longer story. TED Books fills this need.